IN SEARCH OF IRELAND

A permanent end to violence in Ireland and the negotiation of agreed political structures depend on an eventual resolution to even more complex conflicts and confusions of social identity. This book is concerned with the meaning of place in Ireland and with challenging the myths of territory which have been invented to legitimate various constructs of national identity. It argues that the idea of an Ireland divided between North and South, Protestant and Catholic, unionist and nationalist, is a negation of a place that can be more plausibly reinvented as a geographically diverse and socially hybrid world.

Contemporary interpretations of Ireland increasingly stress the diversity of Irish place and society and the fluidity of Irish identity. *In Search of Ireland* explores such variety, seeking to establish representations of place which embrace the hybrid nature of Irishness and pointing to communalities of identity which might admit to pluralistic readings of Irish society. It brings together a number of distinguished contributors, each examining a partic-ular aspect of Ireland's diverse cultural geography and history. These aspects include:

- the plurality of Irelands and the changing constructions of definitions of Irishness;
- the role of class and gender in constructing and complicating traditional nationalist alignments of identity;
- the role of ethnicity in Irish society;
- the invention and imagining of Irish place;
- the political implications of a pluralistic Ireland.

However, no matter how dramatically narratives of culture and place are being renegotiated, many people both inside and outside Ireland continue to define themselves and their conflicts through simple sectarian stereotypes. This book demonstrates the futility and sterility of these representations of Ireland and its peoples.

Brian Graham is Professor of Human Geography, School of Environmental Studies, University of Ulster at Coleraine.

IN SEARCH OF IRELAND

A cultural geography

Edited by Brian Graham

London and New York

First published 1997
by Routledge
11 New Fetter Lane, London EC4P 4EE

Simultaneously published in the USA and Canada
by Routledge
29 West 35th Street, New York, NY 10001

Typeset in Garamond by Routledge
Printed and bound in Great Britain by T.J. International Ltd,
Padstow, Cornwall

British Library Cataloguing in Publication Data
A catalogue record for this book is available from the British Library

Library of Congress Cataloguing in Publication Data
In search of Ireland: a cultural geography / edited by Brian Graham.
Includes bibliographical references and index.
1. Ireland–Historical geography.
2. Northern Ireland–Historical geography.
3. Northern Ireland–Social conditions. 4. National characteristics, Irish.
5. Northern Ireland–Civilization. 6. Irish unification question.
7. Ireland–Social conditions. 8. Ireland–Civilization.
I. Graham, Brian, 1947– .
DA969.I6 1997
306'.09415–dc21
96–52203
CIP

ISBN 0–415–15007–8 (hbk)
ISBN 0–415–15008–6 (pbk)

CONTENTS

ILLUSTRATIONS

FIGURES

TABLES

CONTRIBUTORS

James Anderson Professor of International Development, Department of Geography, University of Newcastle-upon-Tyne

S. J. Connolly Professor of Irish History, School of History, The Queen's University, Belfast

Neville Douglas Senior Lecturer in Geography, School of Geosciences, The Queen's University, Belfast

Patrick J. Duffy Associate Professor of Geography, St Patrick's College, Maynooth, County Kildare

Brian Graham Professor of Human Geography, School of Environmental Studies, University of Ulster at Coleraine

Nuala C. Johnson Lecturer in Geography, School of Geosciences, The Queen's University, Belfast

Catherine Nash Lecturer in Geography, Department of Geography, University of Wales, Lampeter

Michael A. Poole Lecturer in Geography, School of Environmental Studies, University of Ulster at Coleraine

Peter Shirlow Lecturer in Geography, School of Geosciences, The Queen's University, Belfast

William J. Smyth Professor of Geography, Department of Geography and European Studies, University College, Cork

ACKNOWLEDGEMENTS

An extract from Derek Mahon's poem 'Courtyards in Delft', from D. Mahon (1991) *Selected Poems*, is reproduced by permission of Oxford University Press. Extracts from two poems by W.B. Yeats ('Meditation in Time of Civil War and Rebellion' and 'A Vision') are reproduced with permission of A.P. Watt Ltd on behalf of Michael Yeats; and with permission of Simon & Schuster (from 'A Vision', copyright 1937 by W.B. Yeats, copyright renewed 1965 by Bertha Georgie Yeats and Anne Butler Yeats; and from *The Collected Works of W.B. Yeats, Vol. 1: The Poems*, revised and edited by Richard J. Finneran, copyright 1928 by Macmillan Publishing Company, copyright renewed 1956 by Georgie Yeats). A.P. Watt Ltd also kindly grants permission on behalf of the National Trust to include a short extract from 'Ulster 1912' (*The Years Between*) by Rudyard Kipling. Permission is also gratefully acknowledged to reproduce lines from Patrick Kavanagh's 'Stony Grey Soil' and 'From Tarry Flynn' (HarperCollins Publishers), Michael Hartnett's 'A Visit to Castletown House' (The Gallery Press) and John Hewitt's 'Conacre', 'Providence 3' and 'The Return' (The Blackstaff Press).

Every effort has been made to obtain permission to reproduce material contained in this volume. If any proper acknowledgement has not been made, we would invite copyright holders to inform the publisher of the oversight.

PREFACE

The justification for this book rests upon the argument that Ireland's political problems are created by conflicts and confusions of identity. A permanent end to violence and the negotiation of agreed political structures depend on an eventual resolution to these more complex dilemmas. Moreover, the problems of Irish society cannot be depicted solely in terms of religious–political differences, expressed through opposed constructs of nationalism. As is true of any society, all social groups in Ireland draw upon the past to legitimate and validate both their present attitudes and their future aspirations. They do so, however, within a complex geographical mosaic of locality, class and gender. As a study in cultural geography, the book addresses the contested nature of contemporary Irish identity through a consideration of the meanings of place. The themes of the book are placed within the context of the idea that any social reality must be referred to the space, place or region within which it exists. Places are invented, a myth of territory being basic to the construction and legitimation of identity and to the sanctioning of the principles of a society. Thus place is inseparable from concepts such as empowerment, nationalism and cultural hegemony. Societies and localities are interdependent in that social power cannot be conceived without a geographical context; its exercise shapes space which in turn shapes social power. Consequently, myths of identity which seamlessly interweave place and past are widely used to shape identity and to support particular state structures and related political ideologies.

However, as these constructs are socially situated in time, it is likely that myths of identity embody patterns of expedient exclusivity, thereby ensuring that they can be no more than transient representations of any society. Circumstances change and momentarily dominant images of a society, forged in particular epochs for specific purposes, are inevitably challenged or contested along new axes of identity. The past several decades have witnessed a sustained academic attempt to revise representations of Ireland's past and its imagery of place, a questioning of traditional myths which, on the one hand, were erected to justify independence from Britain and provide an origin–legend for the twentieth-century nation-state and, on the other, to legitimate partition. In justifying their own constructs of exclusivity, both nationalist and unionist narratives of identity were heavily dependent on stereotypical images of the Other. The rhetoric of nationalist Ireland and unionist Ulster was careless of class and gender distinction, concentrating instead upon a crude and masculine ethnic division, which conceals a plethora of finely detailed schisms within Ireland's societies. Such traditional renditions of Irishness – or non-Irishness – are of little relevance in an ever more secularised age in which church attendance is declining, traditional attitudes to women are at last being debated, and unemployment and poverty are rife. Moreover, both parts of Ireland are subject to common European Union legislation and to the repercussions of their shared peripherality within increasingly globalised economies.

Contemporary interpretations of Ireland – recognising the sterility of sectarian iconography – are far more likely to be inclusive and open-ended, stressing the diversity of Irish place and society and the fluidity of Irish identity. This book explores such variety, seeking to establish representations of place which embrace the hybrid nature of Irishness and pointing to communalities of identity which might admit to pluralistic readings of Irish society. However, it is recognised that no matter how dramatically representations of culture and place are being renegotiated, some social groups, in the short term at least, are likely to continue in their refusal to accept any intimations of Irish identity, albeit without any clear understanding of the shifting ground of their claimed British identity. The book argues that any resolution to violence depends ultimately upon a cultural and political reinvention of Ireland that can include Ulster unionists who can no longer define themselves in simple sectarian terms, largely through antipathetic representations of a nationalist republicanism that is increasingly no more than an historical stereotype. Equally, the renegotiation of identity in Ireland has immense implications for many northern nationalists, themselves largely self-defined by traditional representations of Irishness.

This book embraces a variety of perspectives and concerns. The various contributors do not subscribe to an agreed political agenda. There is, however, an overall consensus on the need to deconstruct monoliths of exclusive identity in Ireland in favour of narratives of diversity, inclusiveness, hybridity and fluidity – cultural contexts which have to be matched by political flexibility. I would like to express my gratitude to all the contributors who, without exception, met the various deadlines autocratically imposed upon them. My thanks to Kilian McDaid and Michael Murphy for preparing the figures and to Sarah Lloyd, Olivia Eccleshall and their colleagues at Routledge for all their help and support.

Brian Graham
University of Ulster
October 1996

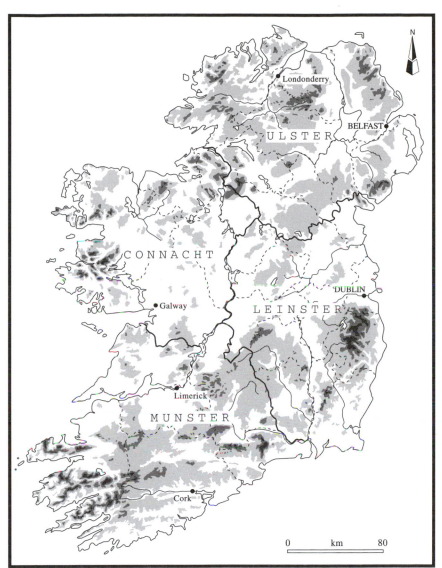

Figure 1. 1 Ireland: provinces and topography

Figure 1. 2 The counties of Ireland

1

IRELAND AND IRISHNESS

Place, culture and identity

Brian Graham

INTRODUCTION

[The country is] like that. Attachment to the soil and aspiration towards
departure. Place of refuge, place of passage. Land of milk and honey and of
blood. Neither paradise nor hell. Purgatory.

(Amin Maalouf, *The Rock of Tanios*)

In 1930, the travel writer, H. V. Morton, published a book called *In Search of
Ireland*. Unlike so many other outsiders, Morton was not beguiled by Ireland
of the Sorrows but concluded his account with the optimistic hope that
'Ireland had emerged from the Celtic twilight into the blaze of day', its
romantic nationalism gone, it was to be hoped, for ever. Nevertheless,
despite such protestations of modernity, his farewell of Ireland, taken on the
Hill of Tara – symbolic centre of the enchanted isle – was redolent of the
stereotypical mysticism that continues to characterise so many renditions of
Ireland and Irishness. For Morton (1930: 273), Ireland was a country yet to
be dehumanised by industrialisation, its typical inhabitant the 'only eternal
figure the world has known; the man who guides a plough'. Ignoring the
industrial heartland of east Ulster only a little to the north of Tara, Morton's
bucolic vision was of an Ireland in which past, present and future might be
harnessed to allow 'a blend of north and south, a mingling of Catholic and
Protestant'. As the millennium approaches, his hopes have yet to be realised.

This present *Search for Ireland* takes place in a very different world. The
'eternal figure' has long departed to the employment exchange and the
suburban housing estate, while his plough has become an artefact in some
museum or heritage centre. Both parts of the island are minor and peripheral
locales within a post-industrial globalised economy, dominated by the new
hegemony of capitalism and its dogma of free trade and competition, by
supra-national political organisations such as the European Union (EU) and
by the unchallenged military power of the United States of America.
Paradoxically, however, beneath this veneer of an ascendant global order,
nationalism and the nation-state survive, together with the bitter, fratricidal
contestations of place that they so often engender. In many consequential

ways, Ireland can be included among the numerous examples of failed nation-state building that litter Europe. On the one hand, people who claim a cultural Irishness live under many political jurisdictions but – most significantly – do so within the island itself. On the other, Irish nationalism has failed to be inclusive of all the inhabitants of the territory which is claimed as congruent with its cultural aspirations. This book approaches contemporary Ireland in the belief that the relationships between the island's geography and culture are fundamental to understanding the confusions and contestations of identity that fracture its peoples. These relationships are equally critical to the formulation of any structures that might lead to an eventual resolution of the conflict.

While the book addresses the dichotomy between North and South, unionist and nationalist, Protestant and Catholic, it is also concerned with several other axes of differentiation that cleave Irish society, even though their significance has often been subsumed within the national conflict. These include class, gender and various manifestations of ethnicity, together with the long-standing but ill-researched schism between urban and rural. In combination with constructs of national identity and spatial patterns of material welfare, such characteristics delineate clearly demarcated social groups. Their implications for identity are rendered more complex, however, by the recognition that any one individual can simultaneously belong to a number of groups. Nationality does not fix class; class does not define gender; gender does not assign ethnicity. Furthermore, the complexity of the social location occupied by any single individual will change through time and vary from circumstance to circumstance. Thus an individual may at one moment be identified as being a Catholic, at another as a woman, elsewhere as middle class, sometimes as Irish, on occasion British and perhaps even European.

The subject-matter may reflect recent ideas put forward within the broad field of cultural geography, but it also responds to the immediacy which these ideas have in any understanding of the axes of conflict and unagreed alignments of identity that split contemporary Irish society. Culture is a notoriously elastic concept but is best regarded here as a signifying system through which 'a social order is communicated, reproduced, experienced and explored' (Williams 1982: 13). It involves the conscious and unconscious processes through which people live in – and make – places, while giving meaning to their lives and communicating that meaning to themselves, each other and to the world beyond (Cosgrove 1993). As Gibbons (1996) forcefully argues, negotiated representations of culture have been at the centre of successive waves of social change in Ireland, not simply reflecting but actively helping to create and transform social experience. While the ideas that will be encountered here are often by no means unique to geography, it is the geographer's concern with place and the often-contested meanings attached to it, which provide the book with its particular focus.

2

PLACE, CULTURE AND IDENTITY

Like any other knowledge, geography is created within specific social, economic and political circumstances. Because it must be situated in this way, the nature of geography is always negotiable, subject to change through time and across space as social and intellectual circumstances alter (Livingstone 1992). Geographical texts and contexts exist in a reciprocal relationship. Regions may have no existence outside the consciousness of geographers 'who, by their eloquence, are able to create place' (Tuan 1991: 693) but, in turn, geographers and their geographies are products of particular social conditions and times. As is the nature of things, these circumstances are unlikely to be agreed, any society being characterised by axes of ideological discord, which may – or may not – be contained by the structures of government and social control.

Contemporary human geography is much concerned with the manipulation of cultural landscape, a complex social construction (Cosgrove 1993) contested along the multiple and overlapping axes of social differentiation. This book addresses four manifestations of the social elaboration of place, all places being imaginary in this sense because they cannot exist for us beyond the socially constructed images which we form of them in our minds (Shurmer-Smith and Hannam 1994: 59). First, it addresses the implications of the notion of contestation of place, most notably the idea that cultural landscapes as allegories of meaning are multivocal and multicultural texts, implicated in the construction of power within a society and capable of being read in a variety of conflicting ways (Cosgrove 1984). These texts interact with social, economic and political institutions and can be regarded as signifying practices 'that are read, not passively, but, as it were, rewritten as they are read' (Barnes and Duncan 1992: 5). Secondly, the concept of socially constructed place is intrinsic to renditions of individual and group identity, which often embody particular readings or narratives of a people's interaction with their cultural landscape. Thirdly, one of the most potent realisations of this process is provided by the formulation of nationalist ideologies, which depend on simplifying synecdoches of particularity, vested in place, in order to summarise and signify very much more complex social structures and to erect criteria of social inclusion and exclusion. Finally, symbolic geographies are also defined by other dimensions of personal and group identity, reflecting the contestation of societies along axes that include class, gender and ethnicity. These, too, are concerned with further criteria of inclusion and exclusion, which interact in complex and diverse ways with nationalistic tropes of identity.

Empowerment and the contestation of places

Because political activity often – if not necessarily – depends on concepts of territoriality, validated through legitimising images of place, landscape texts are frequently central to processes of empowerment. Baker (1992: 4–5) identifies two essential ways in which cultural landscape becomes a framework through which ideologies and discourses can be constructed and contested. First, manipulated depictions of landscape offer an ordered, simplified vision of the world. Secondly, the sacred symbols of a landscape, rich in signs of identity and social codes, act as a system of signification supporting the authority of an ideology and emphasising its holistic character. As Ireland profoundly demonstrates, power cannot be conceived outside a geographical context; social power requires space, its exercise shapes space, and this in turn shapes social power (Harris 1991: 678).

Place therefore forms part of the individual and social practices which people continuously use to transform the natural world into cultural realms of meaning and lived experience. As such, a cultural landscape can be visualised as a powerful medium in expressing feelings, ideas and values, while simultaneously being an arena of political discourse and action in which cultures are continuously reproduced and contested. In one way, landscapes, whether depicted in literature, art, maps and even wall murals, or viewed on the ground, are signifiers of the cultures of those who have made them. They can be regarded as vital texts that mesh with social, economic and political institutions to underpin the coherence of any society. However, because they can be read in different ways by competing social actors involved in the continuous transformation of societies, the meanings attached to these texts remain negotiable. As narratives, they are 'culturally and historically, and sometimes even individually and momentarily, variable' (Barnes and Duncan 1992: 6). Endlessly contested along a multiplicity of dimensions, cultural landscape is thus subject to unrelenting modifications of meaning through time, while remaining an intrinsic quality of the prevailing but transient political economy and its infrastructure of authority. It follows that any landscape signifying the cultural and political values of a dominant group can be viewed as symbolic of oppression by those subservient to – or excluded from – these hegemonic values. For example, the eighteenth-century estate landscapes of Ireland – with their diagnostic triad of features, the demesne, big house and improved town or village – were developed by a largely Anglican *arriviste* landowning élite to place its imprint on an Ireland 'only recently won and insecurely held' (Foster 1988: 192). Later, however, these landscapes came to symbolise English exploitation of Ireland and were excluded from the iconography of the newly ascendant narratives of nationalism that were created in the late nineteenth century.

Identity

If landscape can be depicted as a contested text or narrative, it is clearly implicated in a people's identity, itself embedded in particular intellectual, institutional and temporal contexts. Identity is about discourses of inclusion and exclusion – who qualifies and who does not – and is generally articulated by its contradistinction to a (preferably) hostile Other (Said 1978, 1993). Nationalist identity in Ireland, for example, has been profoundly shaped by presuppositions of malignant Britishness, constructed and presented as 'a collective social fact' that wilfully denies the complexity of that culture (Duncan and Ley 1993: 6). Equally, the very essence of unionism is vested in the assumption of Irish Catholic republicanism as Other, a supposition of timeless uniformity of purpose, people and place that negates the complex diversity of Irish society, geography and history.

These more general ideas can be further refined to focus on the specific context of place, which can be studied in the same socio-psychological manner as concepts such as ethnicity (Entrekin 1991: 54). To Duncan (1990: 17), a cultural – or iconic – landscape is a collage encapsulating a people's image of itself. It symbolises the particularity of territory and a shared past which helps define communal identity, and plays an active part in the reproduction and transformation of any society in time and space. History and heritage – that which we opt to select from the past – are used everywhere to shape these emblematic place identities and support particular political ideologies (Ashworth and Larkham 1994). If it is accepted that the past in this sense is a relative set of contested values, the meanings of which are defined in the present, it follows that a cultural landscape must be negotiable: 'We rewrite history selectively and embed the myth in the landscape' (Smyth 1985: 6). Nevertheless, significant elements of any cultural landscape will be rendered timeless because of the importance – in perceptions of contemporary communality – of deep-rooted continuities with the past which bring about the seeming collapse or foreshortening of time.

Nationalism

Nowhere is this compression of time and space more apparent than in nationalist ideologies and movements which politicise space by treating it as distinctive and historic territory, 'the receptacle of the past in the present, a unique region in which the nation has its homeland' (Anderson 1988: 24). In his influential book, *Imagined Communities: Reflections on the Origins and Spread of Nationalism* (1991), Benedict Anderson argues that any nationalist ideology is the work of the imagination, its communality in large measure self-delusory. It is imagined because the members of even the smallest nation will never know most of their fellow-members, it is imagined as a limited but sovereign entity but, perhaps above all, it is:

imagined as a *community*, because, regardless of the actual inequality and exploitation that might prevail . . . the nation is always conceived as a deep horizontal comradeship. Ultimately it is this fraternity that makes it possible, over the past two centuries, for so many millions of people, not so much to kill, as willingly to die for such limited imaginings.

(Anderson 1991: 7)

But that is not to deny the reality of the discourse, nor the fundamental contribution that significations of place contribute to this perception of social order. Nor are such relationships confined to nationalism alone. Both cultural nation and territorial state claim exclusive sovereign rights over, and access to, territory. Consequently, all states – whether nation-states or not – sponsor intensely territorial official state-ideas. This politicisation of territory is achieved through its treatment as a distinct and historic land, nationalism and the state-idea always looking back in order to look ahead (Agnew 1987: 39–40). The very ubiquity of this relationship between politico-cultural institutions and territoriality suggests that a representation of place is a key component in communal identity, whatever the scale. As one mechanism in the processes that impose homogeneity upon diversity, cultural landscape is fundamental in validating the legitimacy of contemporary structures of authority, structures which are derived, not from the support of a numerical majority alone, but through renditions of plurality – largely fixed in the past – that transcend other social divisions and fix that imagined communality.

Class, gender and ethnicity

Neither group nor individual identity is defined by criteria of nationalism alone. Other criteria of exclusion and inclusion are also implicated in the social construction of a people and its place. In the efforts to impose the homogeneity that constitutes their principal *raison d'être*, however, official renditions of cultural landscape may attempt to elide many of the social complexities emanating from class, gender and ethnicity. These authorised landscapes and places can be viewed as cultural capital – expressions of dominant ideologies (Ashworth 1993) which embody the values and aspirations of that ideology. Officially defined cultural landscapes are therefore directly implicated in the processes which validate and legitimate power structures. Official discourses of place often represent the values of a dominant ethnic group at the expense of minority interests, and promote the interests of social élites while concealing class and gender inequities. However, the very complexity of social divisions in society ensures that even a dominant or hegemonic national landscape (Johnson 1993) may be no more than a transitory representation of place, the ever-present processes of contestation ensuring its continual renegotiation and transformation through time. The parallel existence of other dimensions to identity also produces unofficial representations of place that subvert or challenge state-sponsored nationalism and its narrative of homogeneity.

PLACE, CULTURE AND IDENTITY IN IRELAND

It is apparent, therefore, that identity is defined by a multiplicity of often conflicting and variable criteria. National identity is created in particular social, historical and political contexts and, as such, cannot be interpreted as a fixed entity; rather, it is a situated, socially constructed narrative capable both of being read in conflicting ways at any one time and of being transformed through time. The power of a narrative rests on its ability to evoke the accustomed, a trope that works by appealing to 'our desire to reduce the unfamiliar to the familiar' (Barnes and Duncan 1992: 11–12). The creation of hegemonic landscape narratives facilitates this process by denoting particular places as centres of collective cultural consciousness. As Johnson (1993) argues, the hegemonic image of the West of Ireland as the cultural heartland of the country was an essential component of the late nineteenth-century construction of an Irish nationalism which, in its dependence on a Gaelic iconography, was to prove exclusive rather than inclusive, particularly when its representations became fused with Catholicism. Strongly reinforced by the intellectual élite of early twentieth-century Ireland, the 'West' became an idealised landscape, populated by an idealised people who invoked the representative, exclusive essence of the nation through their Otherness from Britain. The invented, manipulated geography of the West portrayed the unspoilt beauty of landscapes, where the influences of modernity were at their weakest and which evoked the mystic unity of Ireland prior to the chaos of conquest (Johnson 1993: 159). Moreover, this imagery of Otherness was also constructed externally through the conceptualisations of anthropologists, including Arensberg and Kimble (1940). As Agnew (1996) argues, such renditions of place are fundamental to a European tradition of over-simplifying space into idealised constructs of tradition and modernity.

Thus there is little that is conceptually exceptional about the construction of Irish nationalism. The politics of exclusion in nationalist discourse is embedded in all European movements, and contemporary crises of identity are also commonplace. As in Ireland, these often result from the incapacity of the nationalist discourse to assimilate change and resolve the conflicts engendered by exclusion. The creation of the symbolic universe of traditional Irish-Ireland, and its ultimate transformation into a construction of Irishness that was defined by Gaelicism and Catholicism, was 'a supreme imaginative achievement' that began to dissolve only in the 1960s (Lee 1989: 653). By then, it bore no more than a 'tenuous relation to reality' in the South, while it had never accommodated the Protestant, industrialised counties of north-east Ireland. However, Irish-Ireland provided the cultural ethos of the 1937 Constitution, fulfilling the admonition of one nationalist politician that: 'If Ireland as a nation means what [Éamon] de Valera means by it, then Ulster is not part of that nation' (cited in Bowman 1982: 338). The invented geography of Irish-Ireland thus paralleled other dimensions of nationalism to create an Irishness that empowered and legitimised the new state. It was a powerful and exclusive ideology that – particularly through its Catholic ethos – imposed a startling degree of manipulated cultural homogeneity upon the twenty-six counties. For the unionists of Ulster, however, it can be argued that the whole structure of Irish-Ireland became such a powerful expression of Otherness that it was almost sufficient in itself to define Ulster identity. The unionist administration never significantly addressed the cultural vacuum left by partition, preferring instead poorly articulated and even less clearly understood protestations of Britishness (Graham 1994a). Ireland became divided less by the actual border than by the juxtaposition of an increasingly confident Irish identity and a confused and heavily qualified sense of Britishness. It was the former that claimed the moral high ground of legitimacy.

Lee (1989: 653) claims that no alternative self-portrait has yet emerged in the Republic 'to command comparable conviction' with Irish-Ireland. A modernising, increasingly secular state now looks beyond the Otherness of Britain to inclusion within the EU and a markedly more integrated Europe. However, the conflict in the North remains firmly fixed by the polarities of the historical nationalist discourse, ensuring that any future political settlement also demands further renegotiation of identities in both parts of Ireland. Attempts to transform the cultural closure of Irish-Ireland into something more congruent with these contemporary needs for non-exclusivist, outward-oriented, open-ended forms of identity have engendered vitriolic debate, especially among historians and cultural theorists. Although a movement embracing many varieties, the revisionist perspective of Irish history embodies a common stress on the plurality, discontinuity and ambiguity of the Irish past, the antithesis to the narratives of time–space compression – of manipulated homogeneity – which the monolithic Gaelic, rural and, later, Catholic representation of Irish-Ireland imposed on that diversity in the later nineteenth century. Revisionism warns against this retrojection of modern nationalism into the distant past and portrays instead a nation constructed from diverse and often contradictory elements (Brady 1994). The need to take a less Anglocentric view of the past is also seen as being an important part of the revisionist agenda.

Although a useful shorthand label, the notion of a school of historical revisionists, bent on replacing the old nationalist orthodoxies with an alternative framework, is overly simplistic (Brady 1989). As not all revisionists share the same agenda, it is misleading to refer to a revisionist consensus. Disappointingly, the debate has been intensely parochial, receiving relatively little attention outside Ireland (Boyce and O'Day 1996: 7–8). Nor has it had any significant impact on political consciousness within Northern Ireland. Furthermore, the most important implications of revisionism have often been subsumed by accusations that it attempts to rehabilitate the British presence in Ireland, its proponents belittling the suffering and oppression in Ireland's past and acting as apologists for the role of the oppressors (Whelan 1991). Deane contends that revisionism's role is to legitimise partition, and that its advocates defend 'Ulster or British nationalism, thereby switching sides in the dispute while believing themselves to be switching the terms of it' (1991: 102). The entire debate has also been bedevilled by what are, in essence, ultimately futile arguments concerning the objectivity of the professional historian who 'is not a participant in the past' but instead 'an investigator, working by agreed research rules' (Boyce 1996: 234). History, however, is also a situated knowledge, its researchers embedded within their particular intellectual and institutional contexts. The belief that they strive to produce as objective an empirical reflection of the world as possible is at odds with the central assumption in this book that all knowledge is negotiable, contested through time and across space as social and intellectual circumstances alter. Historians' interpretations – like everyone else's – are shaped by the discourses of which they themselves are part.

The linking of revisionism with accusations of objectivity and value-freedom has been unfortunate in that it has unnecessarily obscured the very positive contributions made by historians to the renegotiation of Irishness, while providing neo-traditionalists with ready ammunition. In emphasising the recency of narratives of homogeneity and the seamless integration of virtually all exogenous influences within a supposedly stable and continuous Gaelic identity, many revisionist studies point to the diversity of Ireland's past, one characterised by the meshing and interaction of a variety of cultural influences, conflicts, invasions, colonisations, trade, social contacts and ideas. The nature of Irishness is to be found in the specific delineation of these inputs.

The deconstruction of Irishness into a multicultural and multivocal diversity has many obvious – and as yet unaddressed – implications for unionists in Northern Ireland who have largely been content to define themselves in opposition to the Otherness of Catholic republicanism. It also has manifest connotations for the relationship between both parts of Ireland and Britain, the latter generally treated as collective social fact by unionists and nationalists alike. Gibbons (1996) seeks to accommodate these multiple strands of heterogeneity through the concept of post-colonialism, arguing that Ireland – largely white, Anglophone and Westernised – was paradoxically also a colony within Europe. This past necessitates a present conditioned by post-colonial strategies of mixing and defined by notions of hybridity and syncretism rather than by the 'obsolete ideas of nation, history or indigenous culture'. He writes:

there is no possibility of restoring a pristine, pre-colonial identity: the lack of historical closure [enduring partition] . . . is bound up with a similar incompleteness in the culture itself, so that instead of being based on narrow ideals of racial purity and exclusivism, [Irish] identity is open-ended and heterogeneous. But the important point in all of this is that the retention of the residues of conquest does not necessarily mean subscribing to the values which originally governed them.

(Gibbons 1996: 179)

While the colonial model is an obvious and superficially attractive one, it too offers an unduly stereotypical rendition of the complex negotiations of identity and social interactions that characterise Ireland's past (see, for example, Connolly 1992). However, if the open-ended and heterogeneous qualities of identity alluded to by Gibbons are not merely predicated on *a priori* assumptions of colonialism, his argument shows once again that because human landscapes and other cultural artefacts are defined through the meanings attached to them, they are narratives and allegories that will be renegotiated and transformed as societies are renegotiated.

It is perhaps true to say that Ireland's geographers – if less ready to face the political implications of their findings – have long been aware of the island's regional diversity and cultural heterogeneity (Graham and Proudfoot 1993). In part, this is a reflection of the geographer's interest in the dissimilarity between places, but it is also a function of the markedly Francophile influences on the history of Irish geography, expressed most cogently through the work of Estyn Evans and Tom Jones Hughes. Evans's philosophy, for instance – if heavily influenced in the first instance by the ideas of H.J. Fleure, who held that the 'delineating of regional particularities' was crucial to the evolutionary 'promise of scientific synthesis' (Livingstone 1992: 276) – owed much to the ideas of Paul Vidal de la Blache and, later, to the *Annales* school of *géohistoire* (Graham 1994b). Vidal de la Blache emphasised the significance of ordinary people and their environment: to him, the region was not simply a convenient framework, but a social reality indicative of a harmony between human life and the *milieu* in which it was lived (Claval 1984; Cosgrove 1984). His ideas of place were crucial to the emergence of the *Annales* school, and its concerted attempt to map and explain the complex reality of human life by reference to local and regional studies. It became a tenet of *géohistoire*, particularly as interpreted by Fernand Braudel, that any social reality must be referred to the space, place or region within which it existed. From all this, Evans took the idea of the *pays* as the geographical mediation of synthesis and continuity, the 'product of a people's interaction with their physical environment over centuries' (Baker 1984: 12). These influences meshed with Fleure's theory of regions as places of lived experience to inform Evans's geography of Ireland: regions were 'not just the product of a symbiotic union of people and places' but also the

10

'consequence of the shifting relationships between people and people' (Livingstone 1992: 285). To Evans, geography was 'the common ground between the natural world and cultural history' (Glasscock 1991: 87).

Although their approaches were very different, Evans shared this idea of the cultural landscape as palimpsest, a democratic text recording the history of the undocumented, with one of his most influential contemporaries, Tom Jones Hughes (Whelan 1993: 49). His work, again markedly influenced by French ideas – in this case, those of Pierre Flatrès (Smyth and Whelan 1988) – is much exercised with the layering of time-worlds in hybrid zones and regions of transition. In examining the naming of places in Ireland, Jones Hughes (1970) shows the complex patterns through which territory was claimed, creating a picture of regional diversity that is much at odds with both the monolithic nature of traditional nationalist historiography and the idea that Ulster alone is *the* separate or different region in Ireland.

Geographers have thus been long conscious of the many different scales, axes and dimensions of regional diversity in Ireland, reflecting more accurately perhaps the popular conceptions of place, which often espouse a localism that is at odds with the invented homogeneity of official representations of place, both North and South. As Whelan (1992: 17) argues, there is a

> question of the degree of congruence between a centralised national history, predominantly driven by political imperatives, and the fissiparous diversity of regional histories, where the focus has tended to be more social and economic in character. Such regional perspectives . . . challenge or subvert the centralised orthodoxy, and in this respect proclaim a genuinely pluralistic message.

It is in this context that the now-extensive series of Irish county volumes, published under the Geography Publications imprint, are so valuable in redressing official, monolithic histories and representations of place.

It can therefore be argued that past and present readings of Ireland's human geography have addressed ostensibly modernistic concepts of post-colonialism – notably the idea of open-ended cultural hybridity and the possibility that popular conceptions of place provide a syncretic nexus that might transcend the sectarianism of official versions of Ireland. However, geographers, themselves situated within professional and institutional structures, have often been loath to explore the political connotations of arguments of cultural heterogeneity in Ireland. Although the canon of Irish geography cannot be accused of a complicit post-colonialism that seeks to emulate rather than challenge the standards of the metropolitan centre (Gibbons 1996: 207), the ramifications of its arguments in the dissembling of the caricatures of Otherness that have constituted official versions of Ireland's histories have often not been pursued. It is the purpose of this book to work towards some resolution of this contradiction and to reconsider the geographical perspective on the imbroglio that is modern Ireland.

THEMES AND PERSPECTIVES

If we take the four manifestations of the social construction of place identified above, these can be regarded as providing the central themes that inform the book's content. This is concerned with empowerment and the contestation of place, examining how power structures have defined – or, arguably in the North, failed to define – representations of place that help legitimate and validate dominant ideologies. Because the essential quality of place is defined by meaning, however, it – like any other social construct – can be contested and renegotiated. This is a function of the book's second theme, the concern with the endless transformation and reproduction of the representations of identity that are constructed to underpin official texts of place. Such narratives of hegemonic homogeneity are, however, called into question by resonances of diversity implicit in the deconstruction of place. The third theme is provided by the 'imagined communality' of nationalism, defined through allegories of Otherness, which create sectarian criteria of inclusion and exclusion, and function to legitimate power structures. Finally, the book addresses the parallel existence of other – often elided – dimensions to identity, which help produce popular representations of place that contradict and subvert the official versions of state-imposed ideology.

Within these broad themes, the consequences for Ireland of the deconstruction of traditional structures of identity are examined. In the post-partition South, an officially sanctioned manipulated geography of homogeneity succeeded in subordinating locality, only because of the prevailing uniform Catholic ethos of the state. Strong local senses of places – if muted – did survive, and it can be argued that they have re-emerged, together with other more diverse class, gender and ethnic trajectories of identity, as loci of meaning in today's secularised, materialistic and urbanised society. In the North, the state never addressed the question of a unifying narrative of place, allowing strong local attachments – generally based on territories defined in sectarian terms – to dominate. The result has been cultural incoherence, political impotence and sectarian conflict, once again combined with the elision of other axes of social identity.

Furthermore, there is a marked dissonance between contemporary political structures and their justification and a cultural reading that points to a diversity in Irish place, identity and society, to an Ireland that is indeed defined by cultural hybridity and syncretism. The conceptualisation of an heterogeneous plurality of Irelands challenges partition, the definition of Northern Ireland by what it is not – Republican, Catholic Ireland – and the concept of the Republic as an homogeneous nation-state. While many might regard such outcomes as positive steps toward the deconstruction of two monolithic oppositions, it is most unlikely that the repercussions of hybridity can be accommodated within conventional structures of zero-sum territoriality.

In discussing such issues, the book is divided into four sections, each prefaced by a brief introduction. Part I, 'A multifaceted Ireland', addresses the issues of regional and historical diversity and heterogeneity. It explores the multiplicity of cultural influences that have shaped Ireland's human geography and the ways in which this variety was ideologically suppressed in the linear narratives of nationalist identity that were constructed in the late nineteenth century. This diversity of place is also apparent in the writing of Ireland, the Gaelic discourse being but one, temporarily ascendant, possibility. The national issue it represented again tended to subsume other dimensions to identity. The analysis in Part II, 'Axes of division and integration', discusses class, gender and ethnicity, examining how these subvert and complicate tropes of state-derived nationalism by producing and elaborating popular constructs of identity. In Part III – 'Territory, nationalism and the contestation of identity' – the relationship between representations of place and both official and popular constructs of identity is explored, a discussion that emphasises the marked contrasts between North and South and also the ultimate illogicality of both the union and a united Ireland. Finally, the implications of the book's arguments for structures of territoriality and sovereignty are examined in Part IV, 'Place, identity and politics'.

Conflict in Ireland often seems so deeply entrenched as to be beyond solution. In part, this reflects the immensely powerful trope of nationalist Catholic identity which gave unionists nowhere to go. In turn, they have responded only with a conditional, ambiguous and ill-justified notion of Britishness which can never accommodate the nationalist population of Ulster. The deconstructions of the monolithic representations of nationalist Irishness and unionist Britishness presented in this book point to the renegotiation of Ireland that is a necessary precursor of political change. That latter may in turn prompt cultural reinvention but, ultimately, the legitimacy of any political structure will depend on the acceptance of heterogeneous regional diversity, a rendition of Ireland that demands as much of the South as it does of the North.

ACKNOWLEDGEMENTS

Nuala Johnson, Pete Shirlow and Mike Poole commented on an earlier draft of this chapter. My thanks to them all.

REFERENCES

Agnew, J. A. (1987) *Place and Politics: The Geographical Mediation of State and Society*, Boston: Allen and Unwin.
—— (1996) 'Time into space: the myth of "backward Italy" in modern Europe', *Time and Society* 5, 1: 27–45.
Anderson, B. (1991) *Imagined Communities: Reflections on the Origins and Spread of Nationalism*, revised ed., London: Verso.

Anderson, J. (1988) 'Nationalist ideology and territory', in R. J. Johnston, D. B. Knight and E. Kofman (eds) *Nationalism, Self-Determination and Political Geography*, London: Croom Helm.

Arensberg, C. M. and Kimble, S. T. (1940) *Family and Community in Ireland*, Cambridge, MA: Harvard University Press.

Ashworth, G. J. (1993) *On Tragedy and Renaissance*, Groningen: Geo Pers.

Ashworth, G. J. and Larkham, P. (eds) (1994) *Building a New Heritage: Tourism, Culture and Identity in the New Europe*, London: Routledge.

Baker, A. R. H. (1984) 'Reflections on the relations of historical geography and the *Annales* school of history', in A. R. H. Baker and D. Gregory (eds) *Explorations in Historical Geography*, Cambridge: Cambridge University Press.

—— (1992) 'Introduction: on ideology and landscape', in A. R. H. Baker and G. Biger (eds) *Ideology and Landscape in Historical Perspective: Essays on the Meanings of Some Places in the Past*, Cambridge: Cambridge University Press.

Barnes, T. J. and Duncan, J. S. (eds) (1992) *Writing Worlds: Discourse, Text and Metaphor in the Representation of Landscape*, London: Routledge.

Bowman, J. (1982) *De Valera and the Ulster Question, 1917–1973*, Oxford: Oxford University Press.

Boyce, D. G. (1996) 'Past and present: revisionism and the Northern Ireland Troubles', in D. G. Boyce and A. O'Day (eds) *The Making of Modern Irish History: Revisionism and the Revisionist Controversy*, London: Routledge.

Boyce, D. G. and O'Day, A. (1996) 'Introduction', in D. G. Boyce and A. O'Day (eds) *The Making of Modern Irish History: Revisionism and the Revisionist Controversy*, London: Routledge.

Brady, C. (1989) 'Introduction: historians and losers', in C. Brady (ed.) *Worsted in the Game: Losers in Irish History*, Dublin: Lilliput Press.

—— (ed.) (1994) *Interpreting Irish History: The Debate on Historical Revisionism*, Dublin: Irish Academic Press.

Claval, P. (1984) 'The historical dimension of French geography', *Journal of Historical Geography* 10: 229–45.

Connolly, S. J. (1992) *Religion, Law and Power: The Making of Protestant Ireland 1660–1760*, Oxford: Clarendon.

Cosgrove, D. E. (1984) *Social Formation and Symbolic Landscape*, London: Croom Helm.

—— (1993) *The Palladian Landscape: Geographical Change and its Cultural Representations in Sixteenth-Century Italy*, Leicester: Leicester University Press.

Deane, S. (1991) 'Wherever green is read', in M. Ni Dhonnchadha and T. Dorgan (eds) *Revising the Rising*, Derry: Field Day.

Duncan, J. S. (1990) *The City as Text: The Politics of Landscape Interpretation in the Kandyan Kingdom*, Cambridge: Cambridge University Press.

Duncan, J. and Ley, D. (eds) (1993) *Place/Culture/Representation*, London: Routledge.

Entrekin, J. N. (1991) *The Betweenness of Place: Towards a Geography of Modernity*, London: Macmillan.

Foster, R. F. (1988) *Modern Ireland, 1600–1972*, Harmondsworth: Allen Lane.

Gibbons, L. (1996) *Transformations in Irish Culture*, Cork: Cork University Press/Field Day.

Glasscock, R. E. (1991) 'Obituary: E. Estyn Evans, 1905–1989', *Journal of Historical Geography* 17: 87–91.

Graham, B. J. (1994a) 'No place of the mind: contested Protestant representations of Ulster', *Ecumene* 1, 3: 257–81.

—— (1994b) 'The search for the common ground: Estyn Evans's Ireland', *Transactions Institute of British Geographers* NS 19, 183–201.

Graham, B. J. and Proudfoot, L. J. (1993) 'A perspective on the nature of Irish historical geography', in B. J. Graham and L. J. Proudfoot (eds) *An Historical Geography of Ireland*, London: Academic Press.

Harris, C. (1991) 'Place, modernity and historical geography', *Annals of the Association of American Geographers* 81, 671–83.

Johnson, N. C. (1993) 'Building a nation: an examination of the Irish Gaeltacht Commission Report of 1926', *Journal of Historical Geography* 19: 157–68.

Jones Hughes, T. (1970) '*Town* and *baile* in Irish place-names', in N. Stephens and R. E. Glasscock (eds) *Irish Geographical Studies*, Belfast: Queen's University.

Lee, J. J. (1989) *Ireland, 1912–1985: Politics and Society*, Cambridge: Cambridge University Press.

Livingstone, D. N. (1992) *The Geographical Tradition*, Oxford: Blackwell.

Morton, H. V. (1930) *In Search of Ireland*, London: Methuen.

Said, E. (1978) *Orientalism*, New York: Columbia University Press.

—— (1993) *Culture and Imperialism*, London: Chatto and Windus.

Shurmer-Smith P. and Hannam, K. (1994) *Worlds of Desire, Realms of Power: A Cultural Geography*, London: Arnold.

Smyth W. J. (1985), 'Explorations of place', in J. Lee (ed.) *Ireland: Towards a Sense of Place*, Cork: Cork University Press.

Smyth W. J. and Whelan, K. (eds) (1988) *Common Ground: Essays on the Historical Geography of Ireland*, Cork: Cork University Press.

Tuan, Yi-Fu (1991) 'Language and the making of place: a narrative–descriptive approach', *Annals of the Association of American Geographers* 81: 684–96.

Whelan, K. (1991) 'The recent writing of Irish history', *UCD History Review* 1991: 27–35.

—— (1992) 'The power of place', *Irish Review* 12: 13–20.

—— (1993) 'The bases of regionalism', in P. Ó Drisceoil (ed.) *Culture in Ireland: Regions, Identity and Power*, Belfast: Institute of Irish Studies.

Williams, R. (1982) *The Sociology of Culture*, New York: Schocken Books.

Part I

A MULTIFACETED IRELAND

INTRODUCTION

The three chapters in Part I explore the diversity that lies beneath the narratives of similarity and homogeneity imposed on Ireland and its society by traditional constructs of place and identity. They show how complex differences were reinvented as narratives of continuity and assimilation, while history and the symbolic meanings of place were manipulated to create the new collective identities demanded by the political transformation of the late nineteenth and early twentieth centuries. In Chapter 2, W. J. Smyth demonstrates how 'there have been and are ... many Irelands', reflecting a geographical mosaic that results from the spatially variable impact of numerous conflicting historical orientations and influences. Despite many cultural continuities, he depicts a politically fragmented island that has always been characterised by complex and often faulted cultural strata, engaged in a continuous regional dialectic that created many frontiers and borderlands, of which south Ulster is but one.

In a parallel historical interpretation which adopts a more robust attitude to continuity, S. J. Connolly also identifies, in Chapter 3, a 'kaleidoscope of identities and allegiances' arrayed across a past that is studded with discontinuities. There have been many claimants to an Irish identity, apparent continuities often concealing 'striking changes in content and definition'. He points to the fragile and contingent nature of the political and cultural identities that different groups have created for themselves, but demonstrates how this complex of changing identities was recast in the late nineteenth century as a linear narrative of Irish resistance to English rule. This manipulation of history to create a new collective identity was matched in Ulster by a unionist class alliance that erased 'alternative lines of cleavage and identification'.

Literature and art provide one means of achieving such reinventions of place. In Chapter 4, Patrick Duffy demonstrates how texts are not mirrors to a reality outside themselves, but communicate and reproduce meanings which vary across time and within cultures. Again, he shows that there are many Irelands, of which the stereotypical but very carefully crafted dream

17

world of the West is but one possibility. This idyll of rural arcadia – in reality a world of poverty and mass emigration – bore the brunt of the nationalist myth of homogeneity while successfully excluding the industrial urban landscape of Ulster. Duffy explores alternative representations of Irish place, again pointing to the hybridity of this 'small and diverse island'.

2

A PLURALITY
OF IRELANDS

Regions, societies and mentalities

William J. Smyth

INTRODUCTION

The aim of this chapter is to set aside stereotypical representations of place and explore, instead, the ways in which Ireland's historical plurality has created – and continues to sustain – a complex diversity of regions and localities, each with their own orientations, experiences and mentalities. Since prehistory, the north, and especially the north-east, has had powerful if oscillating links across the narrow straits to Scotland, and beyond to northern England and the Scandinavian world. In contrast, the south-west of the island has looked more to France and Spain, a perspective often shared with the south-east. But this latter region has also always been intimately connected with Wales and the English West Country while, historically, the 'Pale' region around Dublin had a strong Irish Sea focus with powerful linkages along an axis from London to Chester. The far west was more weakly tied into a west European – often Iberian – orbit, but since the late sixteenth century its destiny has increasingly been interlocked with the cities of the eastern seaboard of America (Smyth 1978). Simultaneously, however, the insular qualities of Ireland as place has meant that many different experiences have had to be contained and shared within a narrow, often introverted, ground. The seemingly eternal quality of the dialectic between the northern and southern halves of the island reveals these compacting characteristics. Whether we look at the distribution of megalithic court tombs, Iron Age art, seventeenth-century population patterns or twentieth-century maps of farm size, an old cultural frontier runs across the map, roughly dividing Leath Cuinn (the northern half) from Leath Moga (the southern) (Byrne 1973).

Thus the answer to the simple question, 'Where is Ireland?', cannot be answered in the singular. There have been and there are many Irelands. The crucial point is that Ireland is an island, the size, shape and space relations of which have had a profound influence on the cultural history of its people. As an island's relative location in terms of its relationship and interconnections with other peoples and place changes over time, the presence or absence of an 'insular' mentality is not strictly a product of an island environment *per se*. Depending on the range of cultural conditions, islands may be 'open' and accessible but, conversely, also 'closed' or introverted. There have been critical periods in Ireland's history when the interchange of external and internal contacts appears to have released great energies and renewed regional vitalities. The era of Celtic Christianity, the Norse port-cities, the Anglo-Norman medieval settlement, Ireland's incorporation into the Atlantic world in the seventeenth and eighteenth centuries, and the more recent European engagement since the 1960s, could all be described as phases of opening out. Again, for better or worse, Ireland's external relations have also been profoundly intertwined with those of Britain. Physical closeness but significant cultural distance have been two critical – if misunderstood – features in the often obsessive and confused relations between the peoples of this archipelago. Ideas and peoples have moved to Ireland directly from the continent but, throughout prehistory and history, Britain has frequently acted as a mediator of those influences – European feudalism, the Reformation, democratic and parliamentary procedures and styles of architecture being just four examples of this filtering process.

Conversely, there have been many forces of introversion. Ireland, unlike Britain, was never directly part of the Roman world, although its cultural and political life was profoundly influenced by impulses flowing from the edge of that empire. Thus Ireland experienced a relatively late full-fledged urban life, allowing for the maturation of a complex rural, hierarchical and familiar culture over the space of a thousand years. Likewise, the post-Roman Germanic waves of conquest and settlement did not immediately impinge on Irish society, although there was a substantial and still underestimated Viking contribution to both the overall culture and the regional diversity of the island. Later, only faint echoes of the Renaissance reached these shores and, even in this century, the experiences and destruction of World War II were, with the exception of Belfast, muffled and distant. Clearly, this remoteness engendered by Ireland's insular position has had a fundamental influence on the nature of social relations, making for much intimacy and solidarity, but also for introversion and bitterness. Smaller nations on the European mainland have, on average, land boundaries with two or three neighbours, resulting in much greater interaction. In Ireland, isolation was compounded by marriage and kinship linkages, which reinforced the tightness of the social networks, bringing mutual support but also claustrophobia.

Much of Ireland's regional mosaic springs from the geographically variable impact of these conflicting historical orientations, but other factors are also at work in promoting diversity. No other European country has such a fragmented peripheral arrangement of mountain land all along its borders. This has enriched Ireland with a diversified scenic heritage, but the complicated distribution of massifs presented severe difficulties to would-be conquerors (A. and B. Rees 1989: 118–39). Likewise, richer lowland regions are scattered and fragmented all over the island, facilitating the evolution of strong regional subcultures (see Figure 1.1, p. xiii). In turn the hills and boglands came to serve as territorial bases for local lordships and, with later phases of conquest and colonisation, often became regions of retreat and refuge. The ensuing regional dialectic between peoples of the plains and those of the hills or bogs is a recurring island-wide feature of Irish society, revealing intricate and often faulted cultural strata, as complex as the geological base itself. In such ways, the island's contemporary cultural geography can be seen as the product of innumerable past processes that have created overlapping layers of people and places through complicated interactions of the forces of continuity, assimilation and change.

THE CRYSTALLISATION OF CULTURAL UNITY AND REGIONAL DIVERSITY

Although often used as a synonym for 'Gaelic', the spread of a 'Celtic' culture in Ireland after *c.* 600–500 BC can best be interpreted as a consequence of the invasions of a few waves of iron-wielding warrior élites, who, in establishing themselves as a dominant ruling caste, inherited a long-inhabited and already diversified cultural realm (Raftery 1984). In searching for his vision of Ireland, Estyn Evans (1981) was quite correct in noting that we have a very restricted view of the Irish people if we think of them only as 'Celts', both overlooking the productive mingling of many varieties of settlers in historic times and ignoring the substantial contributions of pre-Celtic peoples from whom the Celts clearly inherited a great deal. Evans saw this representation as an example of imperialist histories, which suggest completely new beginnings at certain periods while disregarding the complex unwritten alterations and adjustments that are very much part of cultural processes. Evans's geographical philosophy insisted upon patterns of cultural continuity in Ireland which reach deep into the prehistoric past, rather than unduly emphasising conquest and change (Graham 1994). He held that the enduring success of Celtic Christianity reflected its ability to assimilate old and new features into a powerful societal synthesis. Particularly crucial in the period from *c.* AD 400 to 1000 was the establishment of a permanent sedentary culture as evidenced by the ring-forts, cashels and *crannógs* and the literal rebaptising of the whole landscape with an array of family and place-names. It was through these latter that the early medieval Celtic élites invested the now permanently settled and bounded places with enduring symbols of their own identities.

Byrne's outstanding analysis (1973) of provincial kingdoms and subkingdoms – the *tuatha* – remains central to our understanding of the ultimate evolution of Celtic systems of territorial organisation, prior to the twelfth-century Anglo-Norman colonisation. He illuminates the layers of peoples and territories as they existed across the island *c.* 900, demonstrating that the marginalisation of former ruling élites was a major feature of Ireland's cultural and political geography during this time (Figure 2. 1). In Leinster, the former kings of Laigin, the Uí Garrchon and Uí Enechglaiss, ended up on the then-remote eastern slopes of the Wicklow Mountains. The Corco Loigde and related peoples lost their control of the inland core of Munster, yet still commanded the southern and south-western coastal and peninsular regions. In Ulster, the Ulaid began their retreat east of Lough Neagh and the River Bann *c.* 450, although they were to retain command of the north-eastern corner of Ireland until the Anglo-Norman conquest and settlement of east Ulster.

Byrne also skilfully locates the arrangement of 'vassal' peoples in strategic buffer lands: the distribution of peoples such as the Airgialla in Ulster, the Loigis in Leinster and the Gailenga and Luigni in Connacht highlights the pivotal position of the key dynasties which dominated all the core regions by *c.* 800. Most of the latter had come to power with the transformation of political structures during and following the Roman occupation of Britain, which had created new alignments along the Irish Sea and within Ireland (MacNiocaill 1972: 1–41). The Uí Néill dynasty was at the heart of these historic transformations, apparently advancing eastwards from its Connacht base in the first centuries AD to command eventually both the rich lands and the symbolic centres. The ecclesiastical capital of Armagh emerged as the significant nucleus of ideology in the Uí Néill drive for the high-kingship of the whole island, symbolically centred on Tara. But as Byrne (1973: 254–74) argues, the idea of high-kingship never became institutionalised and ritualised: from the mid-ninth century onwards, high kings conquered and ruled by force. They claimed the elusive title but did not achieve an effective island-wide government or administration. Whatever long-term institutional possibilities the notion of the high-kingship might have had were rudely shattered by the successive Viking and Anglo-Norman conquests.

West of the Shannon, Connacht was a mosaic of old surviving peoples and newly expanding dynasties. The lowlands of the Moy valley to the north and the rich limestone plains of what is now south Galway, came to be dominated by branches of the Uí Fiachrach. But the future control of the province lay with the branches of the Uí Briúin dynasty, spreading out from their ancient core around Cruachain in Roscommon to establish a powerful territorial lordship that controlled secondary cores around Tuam, Lough Corrib and along the Shannon. In Leinster, the ancient peoples of the Osraige (now County Kilkenny) came to occupy a strategic buffer zone with Munster. Their rulers grew in authority and autonomy at the time of the Viking settlements because of their control of the key river access to the new port-city of Waterford. Leinster had a north–south political structure, with the prehistoric axis of the kingdom pivoting around the ancient hill-fort lands of the mid-Barrow valley. The Uí Dunlainge came to control the Liffey plains and Kildare, while the Uí Chennselaig, protected by the bulky frame of the Blackstairs mountains, dominated the lowland south-eastern core of the island centred on Ferns. To the north-west, the old kingdom of Uí Failge had contracted under constant Uí Néill pressure while, along the ambiguous wetland boundaries with north Munster, reputedly ancient Laigin peoples such as the Éile, the Arada Cliach and the Arada Tíre were ultimately integrated within the flexible overlordship of the Eóganacht over-kingdom of Munster.

Munster appears in the historical record as the least disrupted, most stable, and culturally and politically the most durable of all the early medieval provinces. From c. 600, the powerful Eóganacht dynasty and its various branches controlled the rich plains of Tipperary and east Limerick, together with the Blackwater, Awbeg and Bandon river valleys. In the far west, the Eóganacht Locha Léin dominated the Kerry lowlands north of Killarney. In Munster, therefore, the mainly east–west topographic structure was complemented by a loose Eóganacht hegemony along the valleys, which in turn was matched by strategic criss-cross alignments of related and powerful vassal peoples. The Déisi (the name literally means 'vassal') occupied the lands between Waterford and the Shannon estuary, while the Múscraige extended from north Tipperary to west Cork. In the southern and western coastal regions, a number of ancient but still powerful seafaring peoples seem to have formed a kind of west Munster federation (Ó Corráin 1972: 1–9). They provided the old Munster corridor into the midlands with the founders of monasteries such as Birr and Seirkieran. It was this key group of ecclesiastics who assisted in the realignment of west Munster political allegiances to facilitate the overall control of the province by the Eóganacht of Cashel and Emly (Byrne 1973: 215–20). This rule was almost confederal in character and did not seek the kind of territorial aggrandisement and lordship which characterised both the Uí Néill and Uí Briúin dynasties in the midlands and Connacht (Figure 2. 1).

At the heart of this cultural complex was the kinship system which, if it no longer served to integrate society as a whole, still defined particular groups within it. From c. 950 onwards, the proliferation of Irish surnames began spreading from the key élite families downwards. Distinctive surnames were a boundary-making device, distinguishing the dynastic heirs – including those of professional and ecclesiastical élites – from the disinherited edges of the kin group, who acquired other surnames. Family names became embedded in specific landed properties and functions, creating an intricate mosaic of both small and large territorial lordships. Nevertheless, despite the fragmentation of institutions of rule, administration and ritual within this quasi-hierarchical system of landed organisations, the existence

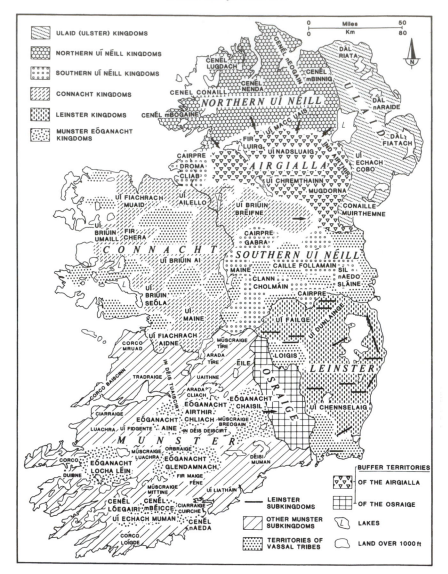

Figure 2. 1 Provincial kingdoms and territories of vassal tribes in
Ireland, *c.* AD 900
Source: Byrne 1973: 120–1, 133, 172–3, 234–5

of named and rooted kinship groups acted as a powerful force in the making
of Irish cultural continuities.

The spread of a universalising Christian religion provided a further dynamic to this society. Its origins may have been located in the south and south-east of the island, reflecting perhaps that region's more intensive interaction with the Roman world of Gaul and south Britain. The original spread of Christianity into Ireland and the evolution of its territorial cores is difficult to understand, for the process was clearly slow, piecemeal and regionally varied (Mytum 1992). In the west, one strong region emerged along an axis from Annaghdown to Mayo. A second embraced the church of Columba and the Ards Peninsula–Strangford Lough region of Down which expanded outwards into the Irish Sea province centred on Iona. The historic core of Leinster churches from Kildare south to Leighlin comprised a third cluster. The most striking feature, however, was the blossoming of a great concentration of monasteries, stretching west from Clonard and Kells to Roscrea and Clonfert. For once, the midlands occupied a core position in Irish culture.

It was the overall wealth of the monasteries and the proto-urban settlements which developed around them that eventually drew the Viking fleets. Driving southwards from the northern sea, Norse power centred on a powerful, fortified settlement at Dublin and its hinterland. The creation of other ports at Wexford, Waterford, Limerick and Cork decisively swung economic and political power away from the Shannon and the midlands and made the control of the southern and eastern coasts and associated seaways central to all Irish futures. But the most decisive Viking contribution – apart from enriching the Irish genetic pool – was to help reorientate the economy and the country outwards. It is also clear that the most successful Irish élites of the period 900–1100 learned much from this engagement with the Vikings, particularly in the consolidation of polities. By the first half of the twelfth century, powerful over-kingdoms – the *mor-tuatha* – were emerging, often spatially coincident with the dioceses demarcated in the twelfth-century reform of the church (Watt 1972). The present-day regional organisation of dioceses was, for the most part, hammered out at this formative time to make for the most enduring of all territorial structures in Ireland's history.

GAELIC AND ANGLO-NORMAN INTERACTIONS AND THE LONG-TERM CULTURAL IMPACT OF THE MIDDLE AGES

Thus early medieval Ireland presents a paradoxical picture of considerable cultural unity, vested in kinship and religious structures, coexisting with political fragmentation. If change and political consolidation was already under way, it is also true that from the late twelfth century onwards the Anglo-Norman colonisation brought settlers and a whole variety of fresh names, innovative ideas about towns, farming and commerce, continental monastic orders and new territorial and governmental structures (Graham 1993). These revitalised, elaborated and deepened existing cultural and trading links within the different parts of Ireland, and between them and the many regions of Atlantic Europe. However, we need to recognise that the application of the ethnic label, 'Anglo-Norman', to all subsequent developments in medieval Ireland obscures as much as it illuminates. Misleadingly, it implies the overwhelming importance of the spread of innovative settlers, instead of recognising the deep interaction between broader, more diverse, currents of European life and the equally varied habitats and societies within Ireland. From the day the first Anglo-Normans landed in south Wexford they were adapting to and learning to survive and prosper in a new land, while facing up to the possibilities and constraints offered by Ireland's location, environment and, above all, the resilience and skills of the populations they had come to conquer. But the latter too were to be altered irrevocably by the confrontation.

For the Anglo-Normans, Ireland's complicated distribution of mountains, hills and boglands brought many enduring difficulties. The complicated border zone of interlaced woods, bogs and lakes that comprised the extensive drumlin and wet clay lands, running across the north midlands and south Ulster, formed one powerful barrier. The great midland bogs and woods also acted as refuges for a resilient Gaelic Irish culture, for the Anglo-Normans did not like the wetlands, and neither did their horses. The most profound cultural environment with which they had to deal, however, was that of the mobile, flexible and pastoral society of the grasslands. Heavily armed, well-drilled and land-hungry, the Normans cut through the country like well-trained beagles, smelling out the good land (Mitchell 1976: 183–91). We should not, though, over-exaggerate their eye for country. Like the later Spanish conquistadors in Latin America, they went straight for the wealthy arable cores of the existing culture, understanding very early the strategic centrality of the great monastic and secular centres of early medieval Ireland. They appreciated, too, the crucial importance of labour supply in a society committed to tillage and the production of grain for the market. The relative success or failure of the Anglo-Normans in a number of frontier regions within the island lay in the varied nature of their adjustments to this cultural world. The rich all-purpose grassland soils were characteristically fragmented and located in different parts of the lowlands, accentuating the importance of these various nuclear zones for regional subcultures. Consequently, one does not fully understand the Norman achievement in Ireland without recognising how regionalised and fragmented its subcultures were (Frame 1981).

The Anglo-Normans – certainly agents of change – were also to borrow much from the rooted and culturally unified, if politically fragmented and decentralised, Gaelic and Norse-Irish populations. While the Anglo-Norman colonisation had a powerful impact in the Dublin–Meath metropolitan core, which must have involved a substantial displacement or integration of the older freeholder farmers and the reorganisation of their territories, elsewhere the colonisers constructed their manorial estates and associated parochial structures within a territorial framework that was, at least in part, already defined before their coming (Graham 1993: 66–8). At every scale, there was much dovetailing of older and newer units, but overarching them all was the first centralised government ever established on the island. This pivoted around the castle and administration in Dublin and the devolved county/shire government, the latter in turn centred on the county court and the person of the sheriff. The beginnings of a county system of territorial administration represented a new and significant addition to the political geography of Ireland. Underpinning these patterns was the further development of towns and urban life, the central fulcrum of the Anglo-Norman colonial process. The urban hierarchy was dominated by the southern and south-eastern ports, all of which – excepting New Ross and Drogheda – were Viking foundations. Between them, the ports controlled the richest grain-producing areas in the whole island, along the Barrow and the Nore in the south-east and the Boyne in the east.

However, the hegemony of Anglo-Norman power was short-lived, the later medieval period witnessing profound island-wide transformations of society, not least the transition from Anglo-Norman to Anglo-Irish. These changes were heralded by the great economic recessions and plagues of the fourteenth century, not to mention the contraction of the highly extended and loosely connected Anglo-Norman frontier under the impact of the so-called 'Gaelic Resurgence' from its anchor points in south-west Munster, Connacht, the midlands, south Leinster and Ulster. Although some Anglo-Norman settlements in the marches were abandoned or shrank in size, few towns or well-established settler communities were overrun or obliterated in the arable regions. Even allowing for the new towns of the sixteenth- and early seventeenth-century plantations, as late as 1660 the urban pattern was still predominantly medieval in distribution and character. Most larger centres still maintained their defensive walls, which marked the boundaries between the relatively privileged and now mainly English-speaking urban societies and the essentially rural worlds beyond their gates (Smyth 1988).

Thus in spite of the apparent disappearance or decline of some of the towns of the early Anglo-Norman frontiers, and the collapse of some settlements in more established areas because of changing economic, political and physical conditions, the medieval urban pattern was for the most part a comparatively enduring one, reflecting the ongoing negotiation of new patterns of social accommodation between the several cultures in Ireland. A modified late medieval and early modern Anglo-Irish society emerged which had obviously adjusted to the ecological possibilities and constraints of the Irish habitat and also to Gaelic Irish kinship organisation. The most striking feature of this society was the expansion and consolidation of the great medieval lordships into essentially autonomous regions. Each of these was dominated by key dynasties who developed their own spheres of administration and – in some cases – their own regional legal codes as well. Thus while the decline of centralised English control in Ireland was one ultimate outcome of the Anglo-Norman era, it was more than matched by the localisation of authority into individual lordships. It was these entities, displaying little obedience to Dublin and London administrations alike, that the Tudor administration of a modernising and centralised English state set out to shatter after 1530.

Turning to the long-term cultural impact of the Anglo-Norman colonisation, late medieval Ireland was clearly divided into a diversity of regions marked by the naming and names of places and people. Again, one can easily over-simplify the picture by stressing the polarities between a predominantly Gaelic Irish subsistence pastoral economy adjoining woodland, bog and mountain edges in the north, midlands and south-west, and the more stratified, densely populated rural village communities of the Anglo-Irish south and east. These latter lived within a more individually based property system, exposed to the market conditions that emanated from the local walled towns and major ports such as Waterford and Drogheda. In between, however, reflecting the social ambiguity of late medieval Ireland, were extensive hybrid cultural zones, largely comprised of the pastoral lands that extended south from Roscommon and Westmeath through north Tipperary and Clare into north Cork.

Underneath this broad framework were the cumulative currents of more intangible cultural interactions, which came to be enshrined in townland, family and Christian names. Jones Hughes (1970, 1984) has shown that at most only 14 per cent of all our townland names – first recorded in the seventeenth century – derive from the medieval settlement. The most critical of these incorporate the suffix, 'town', often combined with an Anglo-Norman surname. These place-names highlight the 'Old English' (as the descendants of the Anglo-Irish were known by then) core of east Leinster in the 1640s (see pp. 45–6). A secondary core area emerges in the Bargy and Forth region of south Wexford. In both these regions, the toponymic dominance of 'town' place-names suggests long-established English linguistic supremacy. Weaker core areas for the 'town' zone are found in County Kilkenny, south Tipperary, Limerick and east Cork, and here it is noticeable that a wide range of other hybrid Norman-Irish place-names also appear. In other regions, including parts of Westmeath, east Connacht and east Leinster, hybrid place-names such as *baile* (bally) suggest an intermingling of Gaelic and Anglo-Irish cultures.

The island-wide distribution of family names (as revealed in seventeenth-century sources) reflects even more faithfully and intimately the finely differentiated cultural geography of the island (Figure 2. 2). For example, the scattering of Old English names points to variations more radical than those revealed by place-names alone. The far greater impact

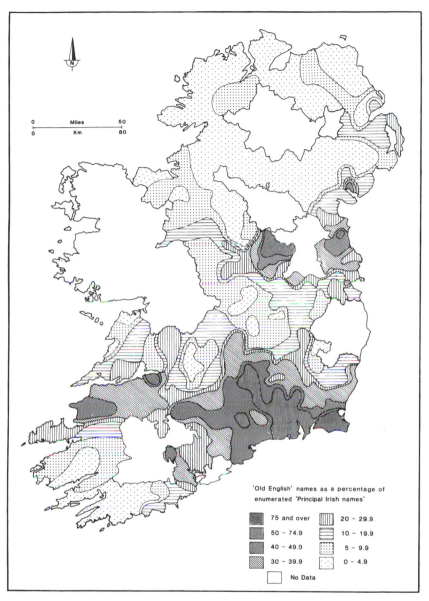

Figure 2. 2 Distribution of 'Old English' surnames recorded in the 1660 poll-tax

of Anglo-Norman culture in the south of the island is here fully revealed, while Old English surnames remained relatively unchanged in the Pale around Dublin. Conversely, in the far west, Anglo-Norman family names were markedly Gaelicised. Overall, the family name distribution strongly reinforces the view that it was in the southern half of the island that the great dialogue between Gaelic and Anglo-Irish was most evenly balanced. It also hints at the importance of the much slower processes of colonisation in the mixed upland and lowland topographies of the south. The rapid expansion of cultural groups along coasts and along the main river valleys did not necessarily involve much mutation of the dominant culture, but the slow, ponderous and piecemeal expansion of settlement overland involved profound adaptations to local exigencies. This was evidently the situation in many parts of Munster and south Leinster. Such processes of adaptation and acculturation were more clearly accentuated in the great pastoral areas of east Connacht and the west midlands.

Even more familiar levels of cultural exchange are revealed when we explore the use of characteristic Christian names among the Anglo-Norman and Gaelic Irish populations (Smyth 1992). Although the island-wide pattern of Christian names cannot be fully reconstructed, the limited evidence suggests that the greatest changes in Gaelic Irish naming patterns occurred in Leinster. Conversely, the most resilient Gaelic Irish naming patterns were in mid- and west Ulster and south-west Munster. And, again, as is often the case in Irish cultural history, east Munster, the Leinster–Ulster borderlands and east Connacht and Clare emerge as the great transitional areas.

Thus the many-sided realities of the Anglo- and Gaelic Irish interactions are revealed. The Anglo-Irish adaptation of elements of Gaelic Irish law and society, as well as their involvement in – and patronage of – Irish language and literature, did not mask a sharp sense of separate identity, particularly in the great lordships of the Anglo-Irish dynasties. Equally, it would appear that these modified Old English cultural worlds exerted a great influence on adjacent Gaelic Irish regions. The interactions and adjustments between the two peoples depended on a wide range of factors – location, the relative size of the two populations locally, the degree of urbanisation and market influence, the nature of the economy and the terrain, and proximity to or distance from major Anglicising centres such as Kilkenny, Waterford and Dublin (Smyth 1985, 1990, 1992). Thus for each region, a different combination of circumstances was at work, all making for a rich and varied mosaic of traditions, naming patterns and dialects.

This hybrid character was most in evidence in the non-material areas of culture – language, dress, literature and sport. Each province, each county, each diocese and even each parish had its own lines of conflict, accommodation and assimilation between the Gaelic Irish and the Old English. Narrower ethnic and political identity came sharply into play when questions of property, legal status, privilege in church and government positions arose. The issue of 'ethnic identity' had more to do, perhaps, with the behaviour and attitudes of the élites than with those of the population generally. The great source of ambivalence among the Old

English landowners lay in their feudal and political relationships with the English Crown, upon which their legal titles to land and other privileges were ultimately dependent. Eventually and fatally compromised by this allegiance, the Old English remained impaled on the cross of a fragmented identity until their material world was shattered, and the basis for their separate identity appropriated, by what T.W. Moody (1976: xliv) has described as 'the most catastrophic and far-reaching changes that took place anywhere in seventeenth-century Europe' – the Cromwellian conquest and settlement.

PLANTATION AND ACCULTURATION: IRELAND AND THE EMERGENCE OF AN ENGLISH-SPEAKING WORLD

Within the broader framework of European cultural history, the Tudor, Cromwellian and Williamite military victories in Ireland during the sixteenth and seventeenth centuries completed a trinity of conquests or half-conquests of this island. This last conquest – that of the 'New English' – was to be the most complete, traumatic and comprehensive, rivalled only by the political, linguistic and institutional supremacy achieved by the Celtic élites and their descendants. A brief examination of several cultural contrasts between Ireland *c.* 1530 and *c.* 1830 reveals the scale and rapidity of the transformations in the political, economic and social geographies of the island over this period. In 1530, Ireland was a land of locally or regionally powerful and relatively autonomous lordships: by the 1830s, administration was efficient and centralised. In 1530, Irish language and literature, along with its patrons and practitioners – the local lords, brehons, poets and clerics – were still in the ascendancy. By the 1830s, the ultimate retreat of the Irish language to the remoter insular–peninsular edges of the western half of the island was but a few decades away. Again, the single religious tradition of the sixteenth century had given way to a plurality of denominations by the nineteenth, each tending to occupy distinctly demarcated geographical areas. The gap between the landscapes of the 1530s and the 1830s was equally vast. While woodland remained a very significant element in the sixteenth-century environment, and agricultural land-uses were more extensive than intensive, by the 1830s the plough and the spade had colonised more arable land than ever before or since in Irish history. The openness of the sixteenth-century landscape had also disappeared. By the 1830s, landscape was as regimented, regularised and reorganised as society.

Class transformations were also marked in both scope and content. The highly compressed yet still elaborately stratified world of the sixteenth century had long since been extended both upwards and downwards. At the top of the early nineteenth-century social hierarchy were the rich privileged worlds of the great landowners and a small number of industrialists and merchants. Below these, a whole galaxy of new middle-class positions had emerged in town and countryside. The skilled artisan classes had also been dramatically enlarged and diversified. At the other end of the class spectrum was the teeming and rapidly expanding impoverished and marginalised mass of the population, subsisting on roadside, bog, mountain and coastal edges or in the cabin suburbs of the cities and towns.

All these changes intersected with a deeply fragmented society, characterised by complex political divisions. The ebb and flow of late sixteenth- and especially seventeenth-century politics in Britain and Ireland transformed society in the island. In the wake of conquest, a new ruling élite, ethnically defined by its origins and religious conformity, took most of the glittering prizes and lived out its dream of ascendancy as the local, if often ambiguous, agent of the wider British state. The landscape was reconstructed around demesnes, big houses and improved towns and villages (Graham and Proudfoot 1994). The losers, defined as different and subservient by the ethnic marker of their Presbyterian or, more particularly, Catholic beliefs – by their non-conformity – became involved in a long and intricate process of assimilation to new legal and cultural norms. Ireland was to remain a deeply divided society which continued to produce a whole series of hyphenated Irish men and women – the Scots-Irish, the Anglo-Irish, the Gaelic Irish and the Catholic-Irish.

I can begin a more detailed evaluation of the changing cultural geography of the island during this period by examining the distribution of immigrant communities in the country by 1660. Possibly as many as 100,000 migrants entered Ireland between the 1590s and 1690s, creating three core areas of settlement. The largest and most enduring was in the north and north-east, and comprised significant Presbyterian as well as Anglican communities. A second Anglican core extended out from Dublin into the south midlands and along the east coast as far as north Wexford. In the long term, the least enduring plantation region was in south Munster. But there were many other regions – the north Leinster plains, extensive if fragmented belts of south Leinster and Munster and particularly much of Connacht – in which planter settlement was not at all significant (see Figure 2. 3).

Both within and beyond the planted regions, perhaps the most critical transformation in early modern Ireland related to the ownership of property in town and countryside alike. In 1600 more than 80 per cent of Ireland was still held by Catholic owners, but by 1641 this figure had been reduced by plantation, intrigue and purchase to 59 per cent (Clarke 1976: 235–7). After the Cromwellian wars and settlement, Catholic ownership declined to 29 per cent while, by 1703, only 14 per cent of the land remained in the hands of the old owners, a residual figure further reduced during the eighteenth century. No other European country witnessed such upheaval in the composition of its landowning élites. However, it should be noted that these processes of displacement and dislocation were not uniform throughout the

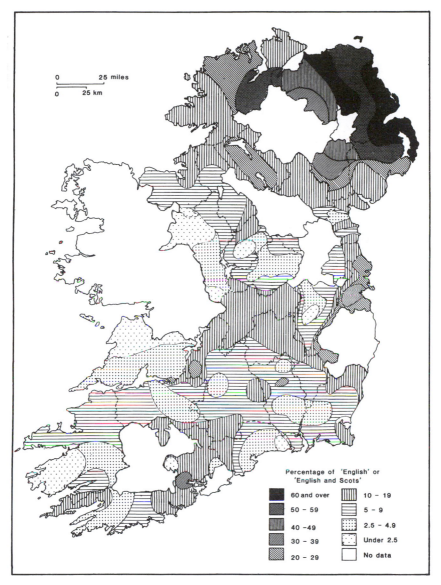

Figure 2. 3 Distribution of 'English' and 'Scots' adult populations recorded in the 1660 poll-tax

island. With some local exceptions, gentry dislocation was a dominant feature of that half of the island north of a line stretching from Drogheda to Ballina. In contrast, much of south Connacht and Clare retained a significant proportion of their old gentry families, as did the Pale region of north

Leinster. Munster and south Leinster reveal a classic dual structure, characterised by a mosaic of old and new élites, while in the cities and towns the processes of displacement were generally more variable. In Dublin and the planted areas, the Old English mercantile class was usually replaced by the new immigrant middle class. Elsewhere, particularly in towns of medieval origin in Connacht, much of Munster and south Leinster, the Catholic merchants survived – and prospered.

This colonial period also witnessed an enormously rapid economic transformation. It is clear that a more commercially oriented market economy had long been a feature of the urbanised lands of the south and east. However, the seventeenth century saw a rapid monetisation and commercialisation of the entire island-wide economy, marked by the equally swift spread of fairs and markets. At the centre of these innovations were the towns, whether reconstructed or – more often – newly built, which must have given much of the seventeenth-century Irish landscape a dramatically novel appearance, above all in those regions so long dominated by essentially rural cultures. This is likely to have been true of Ulster, the midlands, much of Connacht and west Munster and indeed pockets elsewhere, as in Wicklow and north Wexford.

In part, Ireland's growing economic integration, via cities, towns, fairs and markets, was a function of increasing regional specialisation as the Irish economy became more and more subservient to the changing requirements of the British and Atlantic economies. Grain production remained solidly rooted in the old medieval arable cores of east and south Leinster and southeast Munster. Cattle farming, while widespread, became more strongly associated with the west midlands, north Munster and east Connacht, where sheep farming also became conspicuous. Munster was the heartland of the dairying and provision industry, with parts of Ulster also playing a significant role in this sector of the economy, although it was overshadowed by the now rapidly expanding rural-based flax and linen industry.

The development of a landlord culture across the island from the seventeenth century onwards created a new geographical synthesis, clearly demonstrated by the rapidly changing distribution of Irish and English speech. There appear to have been three kinds of core regions in the diffusion of English speech and images: the north-eastern region of Scottish and English immigration and settlement; the Anglo-Irish cultural world centred on Dublin and the midlands core; and the English-speaking worlds of the major port-cities and towns elsewhere in the island. It was from these regions that the battle was waged locally and regionally between an urbanising, imperial, print-based and aggressive English language and culture, and an Irish language and culture that was rural-based, more oral and manuscript-centred. Crucially, it had already gone into decline, both socially and geographically, even before mass communication, high levels of literacy and the forces for democratic social participation effectively developed across the island.

In the 1760s the north-east already stood out both as the region of dominant English speech and also as having the highest levels of literacy on the island. North Leinster and south Ulster was a contracting zone of higher levels of Irish speech and illiteracy in English. It formed a buffer land, under pressure from both the literate north-east and the second growth zone of dominant English speech in the Dublin metropolitan region and much of Leinster. Tipperary and Limerick, together with Kilkenny, formed part of a transitional region, rapidly moving towards English speech-dominance, which extended through the east Connacht borderlands into east Donegal. Conversely, south and west Munster – and even more emphatically the rest of Connacht as well as west Donegal – emerged in the mid-eighteenth century as the great bastions of Irish language, poetry and music.

Adams (1973, 1974) presents a compelling picture of changes in the pattern of language in Ulster. From the north-eastern core of the long-planted Scottish and English settlements, English spread southwards along the Lagan corridor. Adopted as a second language during the eighteenth century, it was established as the primary – and finally as the only – language by c. 1850. The core area of expansion extended south-wards through Fermanagh and Sligo, essentially cutting off the rich literate Irish culture of south-east Ulster from the more oral-based Irish culture of the north-west. Irish-language areas were becoming isolated islands within the island. Even in the strongly Irish-speaking areas of north Donegal, English was beginning to spread from the early nine-teenth century onwards as it followed the routes that carried people and trade (Figure 2. 4).

As with other cultural indicators, Ireland's language patterns reveal a nuanced and complex regional matrix, as well as interesting gender differ-ences, in the rates of language change. What is clear is that by the mid-nineteenth century, Irish had become – excepting some regions of south Munster and much of Connacht – the language of the poor. In the first instance, its areas of strength were confined to the more introverted village clusters of west Connacht. Elsewhere, it was most characteristic of those areas with the lowest land values, the greatest levels of subsistence farming and least affected by the landlord culture with its attendant network of roads and 'improved' towns and villages. The overall pattern suggests that by the early nineteenth century, Ireland was already divided into four different kinds of society: a now receding oral-based Irish-speaking culture; a modernising increasingly literate English-speaking culture, which exhibited an often confused amalgam of both intrusive and indigenous characteristics (Cullen 1981: 18–24); a still triumphant if rather brittle land-centred Anglo-Irish cultural élite, which had already reached the zenith of its achievement; and a burgeoning urban-industrial culture in the north-east, centred on the strongly Presbyterian city of Belfast. This latter culture region had its roots in an older, fully fledged, settler society which had become embedded in what was previously the most Gaelic and most resilient of all the Irish provinces – Ulster.

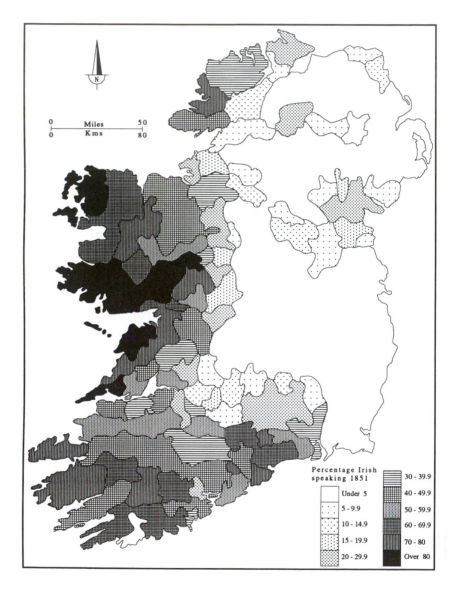

Figure 2. 4 Percentage population speaking Irish in 1851
Source: Census of Ireland 1851

36

DECOLONISATION, FRAGMENTATION AND POLITICAL DIVISION

The late nineteenth-century failure to establish Home Rule for all Ireland culminated in the partitioning of the island in 1920–1. The political and cultural transformations which preceded this rupture added further dimensions to the already heterogeneous cultural geography of the island – indeed, events from the 1780s to the 1820s were decisive in shaping the Irelands of the future. Attempts to create a political movement which would have integrated 'Catholic, Protestant and Dissenter' in a reconstituted and more democratic Irish parliament ended with the Act of Union, which became law on 1 January 1801. Afterwards, direct London-based state intervention in Irish affairs became a dominant feature of Irish life. The creation of island-wide police and national school systems, the establishment of networks of dispensaries, asylums, jails and poor law union workhouses, were all evidence of the state's attempts at social control in the nineteenth century – processes which often involved the main churches (including the Catholic) as allies in this most turbulent of centuries.

The resurgent Catholic church is one of the most potent symbols of this transformation in the cultural, economic and political geography of the island. Whelan (1990) has documented the existence of a core area of the reconstituted Catholic church in the east and south-east, which powered the establishment of Catholicism as a central force in Irish life. The patterning of the nineteenth-century Catholic church reflects an institution that had to seek out and put down parish structures and roots around the newly emerging axes of social and economic life. Ireland was exceptional in Western Europe in that a dual parish system emerged – one Anglican, one Catholic. What this meant was that the Catholic church was in a very good position to adapt to the rhythms and stresses of the emerging Ireland. While very conscious of its medieval roots and territories, flexible chapel- and parish-building strategies allowed it to adjust swiftly to urban and rural population mobility (Figure 2. 5).

This adaptation and reconstruction also saw the modern Catholic church launch an assault on archaic traditions and beliefs, as well as on the looser and more regionally diversified rural worlds that persisted well into the nineteenth century. Thus it adjusted to – and sought to shape – the emerging dominant social formations in Irish society, being part of a more commercialised, hybrid, thrifty, utilitarian English-speaking Ireland, which was expanding its cultural space between, on the one hand, the retreating and introverted Gaelic order and, on the other, an unsympathetic and weakening colonial world. In this pivotal position, the Catholic church became a crucial mobiliser or facilitator of new energies and, in part, a creator of fresh images of Ireland. These greatly assisted the restoration of a feeling of dignity among individuals and groups and renewed a sense of cohesion in many communities, particularly those traumatised by the Famine of the 1840s and the great exodus associated with large-scale emigration, which, by 1890, saw four out of every ten people born in Ireland resident abroad. By the second half of the nineteenth century, an administrative framework of dioceses and parishes – underpinned by a now completed urban and village hierarchy – had been set up throughout the island by a mainly English-speaking Catholic clergy. The church thus constructed an autonomous, self-confident, territorial hierarchy, a powerful instrument of acculturation, social control and group demarcation that stretched downwards through a succession of levels to reach Catholic households and individuals alike. The 'big chapel' was therefore an instrument both of decolonisation and of recolonisation (Figure 2. 5).

The fusion of wider political, agrarian and Catholic agendas culminated in the final major agrarian-political assault on landlordism. In the so-called Land War of the late 1870s and early 1880s, the western counties – heretofore more silent, more remote, more passive – adopted a key leadership role. Crucial to this regional shift in mobilisation and politicisation was the great watershed of the Famine. Every aspect of modern Irish life would have been different if that enormous and deeply traumatic transformation in population numbers, social structures, marriage patterns, beliefs, attitudes and languages spoken had not happened. What the Famine meant for the western counties was that the bitter memories of the horrors of those years would not be repeated in the recessionary and difficult years of 1877–9, when the potato crop was again threatened. Equally relevant, the west had been literally opened up to development in the immediate post-Famine decades through the relatively rapid spread of English speech and literacy, and the increased role of towns as centres of their hinterland communities. Thus it was that the small, struggling cattle-farmers, their kin-connected townspeople and some of the returned emigrants of the west, joined forces with Charles Stewart Parnell's Irish Parliamentary Party finally to defeat landlordism in Ireland. Rather like the experiences of the Old English in the mid-seventeenth century, the material basis of the Anglo-Irish was stripped away, their culture being marginalised in the making of the new Ireland. To paraphrase Standish O'Grady, Protestant Anglo-Ireland, which had once owned Ireland from the centre to the sea, ended up on the edges, stranded between two cultures (Lyons 1979: 180). However, in its going, it was to help light the ambiguous flame of an artistic and linguistic renaissance which, in turn, was to underpin the revolutionary movement for political independence from c. 1900 to c. 1920.

Meanwhile, much of the north pursued a different cultural and political agenda. In 1911, 62 per cent of the 577,000 Anglican or Episcopalian population of the island lived in the province of Ulster. Almost all of the 444,000 Presbyterians lived in the same nine counties. The Ulster ethos was predominantly Presbyterian, industrialisation accentuating this pattern 'since it was the Presbyterians who were most strongly represented among the citizens who built the docks, shipyards and linen mills on which Belfast rose to its precarious prosperity' (Lyons 1979: 24–5). The uncompromising Calvinism of Ulster Presbyterianism had its roots in the seventeenth-century plantations. In the nineteenth century (as today), the battle between liberal

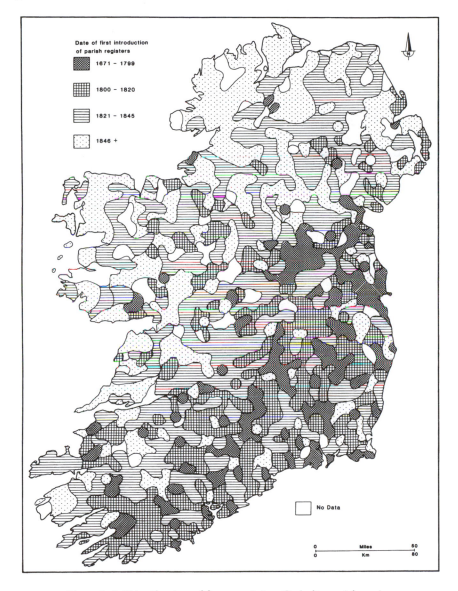

Date of first introduction of parish registers

- 1671 – 1799
- 1800 – 1820
- 1821 – 1845
- 1846 +

No Data

Figure 2. 5 Distribution of first-surviving Catholic parish registers
Source: Based on National Library of Ireland Index

and conservative evangelical wings of Presbyterianism ended in victory for the latter. This helped support the extraordinary religious revival of 1859, which affected all denominations in the north but was primarily an Ulster Presbyterian phenomenon (Gibbon 1975: 44–66).

Again, in the bitter years of the late nineteenth century, the dormant and plebeian Orange Order was revitalised by gentry and industrialists to become more widely representative of the Protestant unionist community, reinforcing a sectarian bigotry which was carried from the countryside into the narrow streets of the industrialising towns and cities. The Orange Order 'fostered a sense of community amongst Protestants and institutionalised the instinct of racial superiority over the conquered Catholics', thus augmenting the high degree of physical segregation which in turn reflected and re-inforced psychic segregation (Lee 1989: 2–3). The political fracturing of the island between 1912 and 1922 nearly led to open war between the two parts of Ireland. These conflict-laden years confirmed northern Protestants in their hatred and fear of southern Republicanism. Northern Catholics were in turn locked into a position of non-co-operation and political inferiority. 'In both communities the already entrenched siege mentality was still further inten-sified' (Lyons 1979: 104–5). There thus developed a lack of congruence between the geographical reach of the new southern state and the wider geographical distribution of a population feeling a sense of belonging to this nationalist community; for unionists the problem has been that their distinctive sense of identity fails to extend even as far as the state boundary (see Chapter 10).

CONCLUSION

The period from *c.* 1850 to *c.* 1950 could be described as a period of necessary 'closure' for much of what became the Irish Free State as the struggle for polit-ical and cultural independence was both intensified and achieved. The striking stability of post-Famine southern society and its norms – determined especially by the ethos of the strong farmer, which in turn was transferred into the Catholic church and the towns – was built upon enormously high levels of emigration from the country as a whole. It was this kind of conservative society which put in place many of the essential institutions of an indepen-dent state and did so with some courage, ingenuity and tenacity. But there were severe and inevitable limitations to the ideology, imagination and capac-ities of the young state, cruelly exposed during the 1950s in a decade of fundamental adjustment and high emigration. Huge changes in the nature of the southern state and society took place in the 1960s as new economic, polit-ical and cultural policies were adopted, not least in terms of industrialisation, the massive development of second- and third-level educational institutions and the radical transformation in levels of information and media influences.

The Republic of Ireland in the late twentieth century is a more complex, fragmented and secularising cultural world.

In part, the imaginative failure to assimilate the north to the early nationalist agendas reflects a failure to recognise that 'much history revolves around fears and prophecies' – especially the northern apprehensions that Home Rule would mean Rome Rule (Lee 1989: 19). But at the beginning of the twentieth century, northern resistance also reflected unionist determination to stay within the wider golden orbit of the then British Empire, so that the northern economy could benefit from the enlarged economies of scale. Since the United Kingdom and Ireland both joined the then European Community in 1973, the wheel has come full circle. In recent decades, it is the Republic which has built and produced, however painfully, slowly and fitfully, the key state institutions and personnel to engage on an equal footing with the countries and markets of an enlarged European and world community. In Northern Ireland, deindustrialisation, the expansion in miniature of a whole range of state functions, and Direct Rule from Westminster, have all made for a more dependent and provincial culture. The paramilitaries on both sides are scarcely forces of liberation, often appearing as reactionary voices and actors, locked into battles which belong more to the late nineteenth than the late twentieth century. The ultimate 'Republican' goal is to reverse both the Anglo-Norman and Tudor conquests. Conversely, for the 'Loyalist' community, the maintenance and assertion of their continued sense of superiority necessitates the retention of a British state apparatus and the non-recognition of the material and symbolic rights of the nationalist community. Neither of these two positions is tenable, feasible or desirable in the context of the complexities, nuances and faultlines of the many Irelands that have existed – and still exist – on this small and diverse island.

REFERENCES

Adams, B. G. (1973) 'Language in Ulster, 1820–1850', *Ulster Folklife* 19: 50–5.
—— (1974) 'The 1851 language Census in the north of Ireland', *Ulster Folklife* 20: 65–70.
Butlin, R. A. (ed.) (1977) *The Development of the Irish Town*, London: Croom Helm.
Byrne, F. J. (1973) *Irish Kings and High-Kings*, London: Batsford.
Clarke, A. (1976) 'Selling royal favours, 1624–32', in T. W. Moody, F. X. Martin and F. J. Byrne (eds) *A New History of Ireland*, III, *Early Modern Ireland, 1534–1691*, Oxford: Oxford University Press.
Cullen, L. M. (1981) *The Emergence of Modern Ireland, 1600–1900*, London: Batsford.
Evans, E. E. (1981) *The Personality of Ireland*, 2nd ed., Belfast: Blackstaff Press.
Frame, R. (1981) *Colonial Ireland, 1169–1369*, Dublin: Helicon.
Gibbon, P. (1975) *The Origins of Ulster Unionism*, Manchester: Manchester University Press.
Graham, B. J. (1993) 'The high Middle Ages: c. 1100–1350', in B. J. Graham and L. J. Proudfoot (eds) *An Historical Geography of Ireland*, London: Academic Press.

—— (1994) 'The search for the common ground: Estyn Evans's Ireland', *Transactions Institute of British Geographers* NS 19: 183–201.

Graham, B. J. and Proudfoot, L. J. (1994) *Urban Improvement in Provincial Ireland, 1700–1840*, Athlone: Group for the Study of Irish Historic Settlement.

Jones Hughes, T. (1970) 'Town and *baile* in Irish place-names', in N. Stephens and R. E. Glasscock (eds) *Irish Geographical Studies in Honour of E. Estyn Evans*, Belfast: Queen's University.

—— (1984) 'Historical geography of Ireland from *c*. 1700', in G. L. Herries Davies (ed.) *Irish Geography Golden Jubilee, 1934–1984*, Dublin: Geographical Society of Ireland.

Lee, J. J. (1989) *Ireland 1912–1985: Politics and Society*, Cambridge: Cambridge University Press.

Lyons, F. S. L. (1979) *Culture and Anarchy in Ireland, 1890–1939*, Oxford: Clarendon.

MacNiocaill, G. (1972) *Ireland Before the Vikings*, Dublin: Gill and Macmillan.

Mitchell, F. (1976) *The Irish Landscape*, London: Collins.

Moody, T. W. (1976) 'Introduction', in T. W. Moody, F. X. Martin and F. J. Byrne (eds) *A New History of Ireland*, III, *Early Modern Ireland, 1534–1691*, Oxford: Oxford University Press.

Mytum, H. (1992) *The Origins of Early Christian Ireland*, London: Routledge.

Ó Corráin, D. (1972) *Ireland Before the Normans*, Dublin: Gill and Macmillan.

Raftery, B. (1984) *La Tène in Ireland*, Marburg: Sonderband.

Rees, A. and B. (1989) *Celtic Heritage*, London: Thames and Hudson.

Smyth, W. J. (1978) 'The western isle of Ireland and the eastern seaboard of America – England's first frontiers', *Irish Geography* 9: 1–22.

—— (1985) 'Property, patronage and population – reconstructing the human geography of mid-seventeenth-century County Tipperary', in W. Nolan (ed.) *Tipperary: History and Society*, Dublin: Geography Publications.

—— (1988) 'Society and settlement in seventeenth- century Ireland – the relevance of the "1659 Census" ', in W. J. Smyth and K. Whelan (eds) *Common Ground: Essays on the Historical Geography of Ireland*, Cork: Cork University Press.

—— (1990) 'Territorial, social and settlement hierarchies in seventeenth-century County Kilkenny' in W. Nolan and K. Whelan (eds) *Kilkenny: History and Society*, Dublin: Geography Publications.

—— (1992) 'Making the documents of conquest speak: the transformation of property, society and settlement in seventeenth-century Counties Tipperary and Kilkenny', in P. Gulliver and M. Silverman (eds) *Approaching the Past: Historical Anthropology Through Irish Case-Studies*, New York: Columbia University Press.

Watt, J. A. (1972) *The Church in Medieval Ireland*, Dublin: Gill and Macmillan.

Whelan, K. (1990) 'Catholic mobilisation 1750–1850', in P. Bergeron and L. M. Cullen (eds) *Comparative Aspects of Politicisation in Ireland and France*, Paris: Seuil.

3

CULTURE, IDENTITY AND TRADITION

Changing definitions of Irishness

S. J. Connolly

INTRODUCTION

Notions of culture, identity and tradition have a well-established place in Irish political debate. The emergence of movements of cultural defence – the Gaelic Athletic Association (GAA), the Gaelic League and the Irish-Ireland movement – were an important part of the 'new nationalism' of the years before World War I. The fostering of a distinctive cultural identity was duly enshrined as a central objective of the new state that was created following the upheavals of 1916–23. In contemporary Northern Ireland, culture remains a live political issue. Irish-language street names, airtime for Irish-language broadcasting, and state funding for Irish-language schools have been visible symbols of nationalist political advance, resisted as such by unionist opponents.

Ideas of tradition have also played a central role in attempts to analyse the current Northern Ireland conflict from outside. The late F. S. L. Lyons, in a study ostensibly concerned with the period before 1939, but the tone and preoccupations of which were clearly much influenced by the then recent resurgence of Northern political violence, argued that Ireland's contentious past should be understood in terms of 'the collision within a small and intimate island of seemingly irreconcilable cultures . . . a diversity of ways of life which are deeply embedded in the past and of which the much advertised political differences are but the outward and visible sign' (Lyons 1979: 177). Four years later a more explicit attempt by academics and others to respond to what was by then the fully developed Northern Irish conflict took as its starting-point, 'the relevance to our problems of violence and civil strife of the existence of two distinctive traditions, cultures and communities' (Two Traditions Group 1983: 3). More recently, the concept of a conflict founded on rival historical and cultural traditions has helped to shape educational policy, inspiring the inclusion of cultural heritage and education for mutual understanding as compulsory elements within the Northern Ireland school curriculum (Lambkin 1996: 65–94).

For the Irish historian this sense of the relevance of tradition and culture

43

to contemporary political problems might at first sight seem to present an enviable opportunity. Works of academic history have found their way onto the Irish bestseller list. Historical debates have been covered by press, radio and television. There is a thriving market for heritage centres and summer schools. Local history societies proliferate. In reality, however, unprecedented public interest in the historian's wares has proved to be a mixed blessing. Those who have tried to respond to this demand have, more often than not, found themselves unable to meet expectations. Consumer dissatisfaction is reflected in the campaign against 'revisionism': originally a reaction against the irreverent tone adopted by some younger writers towards prominent nationalist myths and icons, this has more recently broadened into a general attack on the failure of academic historians as a group to provide a version of the Irish past to which the wider public can relate (Brady 1994). Equally significant has been the decision to bypass the historians altogether: works like *The Field Day Anthology of Irish Writing* (Deane 1991) and Declan Kiberd's *Inventing Ireland* (1995) are ambitious attempts by cultural theorists and literary scholars to construct their own accounts of Ireland's past to set against their analyses of its present.

That Irish historians should have failed to meet the demands made on them by a public debate, in which history is seen as relevant to present problems, is in some ways predictable. A degree of tension is probably inevitable between the academic specialist, concerned to define, document and qualify, and a wider public hoping to find a coherent and accessible overall picture. There is also the inescapable connection between the popularity of Irish history and its controversial character. Precisely because the Irish past is seen as relevant to the Irish present, control of its interpretation becomes worth contesting. But perhaps the most important reason why Irish historians have so clearly failed to meet the expectations of a wide section of their would-be public lies, not in their professional scruples, but in the nature of Irish history itself. A well-established tradition of cultural analysis and political explanation points to continuity and tradition as keys to an understanding of modern Ireland. Yet neither theme plays all that prominent a part in the writing of historians themselves. Instead, what emerges most strongly from a wide range of historical writings are the sharp discontinuities that stud the Irish past, and the fragile and contingent nature of the political and cultural identities that different groups have created for themselves.

MEDIEVAL AND EARLY MODERN IDENTITIES

For the inhabitants of medieval Ireland, definitions of national identity were from an early stage a source of concern, puzzlement and controversy (Frame 1993). The claim that the descendants of Anglo-Norman settlers quickly came to be 'more Irish than the Irish themselves' is a myth. The colonists who established themselves in Ireland from the twelfth century called themselves 'English' rather than 'Norman' or even 'Anglo-Norman'. The much-quoted phrase, '*hiberniores ipsis hibernis*', cannot be traced further back than the seventeenth century – its coinage at that stage, and its subsequent prominence in historical folk-memory, are themselves testimony to the extent to which later phases in the development of Irish ideas of national identity were to depend on an imaginative transcendence of the facts of ethnic origin and past political allegiance (Cosgrove 1979). In place of any simple one-way process of assimilation, recent historiography emphasises the complex and multidirectional nature of cultural inter-change. Families of English descent did indeed adopt Irish language, dress and customs. But Gaelic Ireland was also penetrated by English legal, administrative, economic and cultural forms, and persons of Irish descent were absorbed into Anglo-Irish society. Meanwhile, 'the English born in Ireland' found themselves forced repeatedly to defend local institutions and privileges against attempted encroachments from the other side of the Irish Sea. But there is disagreement on whether, or how far, such conflicts amounted to the emergence of a distinct Anglo-Irish political identity (Ellis 1986).

Into this already complex blend of identities were introduced the new lines of division created by the failure of the Reformation – for reasons which continue to be debated – to win the support of a significant proportion of either the Irish or the English inhabitants of Ireland. By the early seventeenth century, contemporaries had begun to speak of the descendants of the pre-Reformation colonists as the 'Old English'. While this label acknowledged their separateness from the Gaelic Irish, it also distinguished them from the growing number of more recent settlers who were equally enthusiastic proponents of English government and the containment of Gaelic Irish rebelliousness, but who were also – unlike the Old English – supporters of the Protestant Reformation (Clarke 1978).

This tripartite division into Old English, New English and Gaelic Irish, combining ethnic and religious elements, has a pleasing symmetry that has helped to make it a standard feature of historical accounts of early modern Ireland. Yet its neatness is to some extent deceptive. What was considered the Old English community actually included families of Gaelic descent that had been absorbed, culturally and politically, into the English of Ireland. A study of the Catholic Confederation of the 1640s has discovered among its members not just Gaelic Irish and Old English, but also a number of more recent settlers who had either always been Catholic or had become so since their arrival (Cregan 1995: 491–4). Conversely, James Butler, Duke of Ormond, was in political terms New English, a committed defender of the Protestant interest, yet he was also head of a family established in Ireland from the thirteenth century. In addition, numerous connections were formed across both ethnic and religious lines of division by intermarriage.

What these contradictions reflect is the extent to which the labels Old English, New English and Gaelic Irish were themselves the product of a political process of labelling and self-definition. To emphasise the division between Old and New English served the purposes of Elizabethan and later settlers who had been attracted to Ireland by the military and administrative openings, and the opportunities for personal enrichment, created by the wars and rebellions that accompanied the more complete absorption of Ireland into the Tudor state. But real gains could only be made at the expense of existing interests. The dispossession of the Gaelic Irish could be legitimised on grounds of culture, religion and political allegiance. The Old English, by contrast, could be attacked only on one front – religion. A label that set them apart on this basis from more recent settlers could thus hardly fail to be attractive. To the self-interest of enemies, moreover, was added the solicitation of would-be friends. The late sixteenth and early seventeenth centuries saw the development of a new sense of Irish Catholic identity. Originating among Irish *émigrés* in the universities and armies of Catholic Europe, and strongly influenced by the militant spirit of the Counter-Reformation, this sought to divert attention from the ethnic and cultural barriers that had for centuries divided the Gaelic Irish and the English of Ireland, emphasising instead the new bond of a shared religious allegiance (Cunningham 1986a: 165–70). The redesignation of pre-Reformation settlers as the Old English, an established part of Irish society, helped to smooth over centuries of warfare and mutual hostility. The invention about this time of the pseudo-medieval formula, 'more Irish than the Irish themselves', was part of the same process. So too was the production of what was to become one of the founding texts of Irish historical writing, Geoffrey Keating's *Foras Feasa ar Eireann*. Written in Irish by a priest of Old English descent, this created an elaborate composite of existing origin-legends in order to present the English invasion of the twelfth century as only the latest in a series of episodes of conquest, settlement and cultural assimilation (Cunningham 1986b).

THE TRANSFORMATION OF THE SEVENTEENTH CENTURY

The position of the Old English, caught between a hostile Protestant establishment and the growing militancy of the Counter-Reformation, was precarious from the start. By the mid-seventeenth century, it had become impossible. In retrospect the defining moment came in 1641 – at the start of the civil wars that engulfed all three kingdoms of the British Isles – when the Old English and Irish entered into a political and military alliance as the Confederate Catholics of Ireland. Yet long-standing differences of culture, self-image and political allegiance did not disappear overnight. The Old English took up arms in 1641 in the belief that the monarchy was about to lose power to a strongly Protestant English parliament. They insisted that this was a loyal rising, in defence of the royal prerogative. Their Irish allies, by contrast, were more ready to see themselves as an arm of the international forces of the Counter-Reformation and, to some extent, as defenders of Ireland against foreign rule. When the desire of the Old English to reach a new accommodation with the monarchy came into conflict with the priorities laid down by the papal representative sent to advise the Confederation, there were bitter conflicts, leading at one stage to military confrontation. None of this, however, saved the Old English from being categorised along with the Gaelic Irish as popish rebels, or from being included in the massive confiscations of landed property that followed their defeat.

The civil wars of the mid-seventeenth century also brought to prominence another element in Ireland's complex mix of ethnic and religious identities. These were the Scots who had moved into Ulster, through a combination of formal plantation schemes and spontaneous migration, over the previous half-century. They brought with them a distinctive culture, reflected in speech, dress and popular custom, as well as a different strain of British Protestantism. In the civil wars of 1641–53, the Scots of Ulster, reinforced by an army from Scotland itself, played a separate and distinctive role, at different times aligning themselves with and against each of the other major groupings – Protestant Royalists, Protestant supporters of parliament, and even Confederate Catholics. Meanwhile their separate religious identity, up to then uneasily contained within the episcopal Church of Ireland by a tacit mutual accommodation, was formalised in the establishment of Presbyterian congregations. From 1660, after the restoration of monarchy and episcopacy, Ulster Presbyterians began a long period of indeterminate status. As Protestants they enjoyed a degree of acceptance; from 1672 their ministers even received an annual state subsidy, the *regium donum*. Yet as religious dissenters they were also exposed to periodic harassment, when meeting houses were closed down and ministers silenced or arrested. In addition, government remained deeply suspicious of their continued Scottish links, and every outbreak of political or religious disturbance in Scotland during the 1660s and 1670s saw extra troops moved into Presbyterian areas of Ulster.

The distinction between Old English and Irish persisted in muted form into the second half of the seventeenth century. The conflict between allegiance to papacy and to crown reappeared in the 1660s, again with a strong ethnic dimension, over proposals for a loyal Remonstrance to be presented to the restored Charles II. Some studies also propose a continuing division in religious culture. The Old English, it is suggested, responded more enthusiastically to the reformed Catholicism, purged of folkloric accretions and organised within a firm territorial framework of diocese and parish, defined

47

by the Council of Trent. Gaelic Ireland, by contrast, remained more attached to traditional forms. Some Protestant observers in the Restoration period were still willing to distinguish between the native Irish and the less dangerous, and possibly even less uniformly disloyal, Catholics of English descent. But the significance of such distinctions was beginning to fade, particularly now that both Old English and Irish were perceived as having a common interest in reversing the mid-century land confiscations.

The Old English gentry and aristocracy were still disproportionately represented in the Catholicised military and civilian administration built up by the Earl of Tyrconnell, following the succession of James II. An ethnic dimension can also be detected in the divisions that emerged, following the first defeats in the war of 1689–91, between advocates of continued resistance and those who favoured a negotiated settlement. Such distinctions, however, were ignored by opponents. Contemporary Williamite accounts of the war of 1689–91 generally referred to the Jacobite army indiscriminately as 'the Irish' and to their own forces, even when these were partly or wholly composed of Irish Protestants, as 'the English'. The eventual Williamite victory, and the subsequent enactment of penal laws formalising the inferior status of Catholics as a whole, confirmed the irrelevance of earlier ethnic divisions. Old English and Gaelic Irish were now united in irreversible defeat, with none of the apparent possibilities for differential treatment that had lain behind earlier conflicts of interest (Connolly 1992: 114–24).

THE EIGHTEENTH-CENTURY CONTESTATION OF IRISHNESS

At the same time as long-standing uncertainties concerning the interplay between religion and ethnicity among Irish Catholics were thus being resolved, important changes had begun to take place in the self-image of the third major group within Irish society. By the beginning of the eighteenth century, the idea that Irish Protestants were Englishmen living on the west shore of the Irish Sea had already begun to lose conviction. Most 'New English' families had been settled in Ireland for fifty or sixty years, many for up to a century longer. The development of a new sense of identity was also encouraged by political discontents. Ireland's subordination to England (from 1707 Great Britain) had from an early stage given rise to recurrent conflict. But the transfer of power from crown to parliament, accomplished in the British revolution of 1688, had sharply reduced the mediation through a common monarch that had earlier helped to make subordination tolerable. Meanwhile, the rhetoric of parliamentary constitutionalism that had accompanied the revolution, helped to sharpen awareness of the inferior status of the Irish parliament. There were also more concrete resentments. Restrictions on trade and manufacturing industry imposed to protect British interests from Irish competition were widely, even if in the judgement of modern economic historians mistakenly, blamed for the depressed state of the Irish economy. The use of positions in the Irish church, judiciary, army and civil administration as rewards within the British system of political patronage reduced the share of seats available to natives. There was also the growing awareness that in Britain itself Irish Protestants were no longer seen as fellow Englishmen. A revealing development of the early eighteenth century was the appearance in the English theatre of a new figure of fun. To the two existing varieties of stage Irishman, the half-savage Catholic fanatic and the comically slow-witted servant, was now added a third stereotype. This was the Irish landed gentleman, by implication a Protestant, but displaying an élite version of the traditional Irish vices in his addiction to duelling and gambling, his incessant hunt for a well-fortuned bride and his combination of a shaky fortune with ludicrous pride in an inflated family history (Hayton 1988).

The patriotism of eighteenth-century Irish Protestants was to be outrageously romanticised by later generations. It provided a precedent for nationalist claims, while to summon up the memory of Swift and Grattan was also an invaluable means of undercutting contemporary Protestant unionism. Historians, partly in reaction, have been cautious in their assessment. Irish Protestants, it has been argued, did not embrace Irishness; it was forced on them by a hostile or indifferent British government (Smyth 1993). Their patriotism was pragmatic and episodic. They made no pretence of defending the rights of a nation: rather, like contemporary defenders of corporate or provincial liberties in France and elsewhere, their commitment was to a particular community, the Protestant propertied classes, and its distinctive institutions (Leerssen 1988). Despite all this, it remains the case that it was almost exclusively from the Protestant middle and upper classes of the eighteenth century that a claim to Irish political autonomy was first systematically articulated. In doing so they developed political models, rhetoric and imagery that were to continue – right up to 1914 – to shape the aspirations of the great majority of Irish nationalists.

Nor should we underestimate the emotional power that patriot language and symbolism had, by the second half of the eighteenth century at least, developed. As early as the 1720s, there had been signs of a more positive attitude towards aspects of Ireland's history and antiquities. In the exhilaration surrounding the successful patriot agitation of 1778–82, which secured the removal first of commercial restraints then of restrictions on the legislative autonomy of the Irish parliament, enthusiasm for all things Gaelic reached new heights. The Royal Irish Academy was established in 1785, with the Earl of Charlemont, Commander-in-Chief of the Volunteer movement that had played a key part in the achievement of legislative independence, as its first president. Four years later, Charlotte Brooke's *Reliques of Irish Poetry* inaugurated what was to be a long-lived tradition of Anglo-Irish verse drawing aesthetic and political inspiration from a real or imagined Gaelic past (Vance 1981).

The transformation of the descendants of New English colonists and Cromwellian warriors into fervent Irish patriots is at first sight the most striking of all the redefinitions of self and others that have taken place in Ireland's complex history. If the conversion was encouraged by residence over several generations, allied to dissatisfaction with the workings of the Anglo-Irish connection, it was also made possible by the almost complete disappearance of Gaelic Ireland. Irish remained the language of the majority of the population. But in other respects – for example, dress, diet, economic behaviour and family relationships – the early and mid-seventeenth century had seen a major reshaping of popular culture, transforming the Irish lower classes from a self-evidently alien race into what could be seen as a poorer, less orderly counterpart of the English peasantry. Meanwhile, a series of defeats and land confiscations had destroyed the web of patronage, clientship and economic dependency on which the power of the Gaelic ruling class had rested. Irishness, defined in terms of the Gaelic tradition, had thus become a vacant property to be appropriated in the service of current political needs. The obvious contrast here is with Scotland, where the military and political structures of Highland society remained intact for more than a century longer, and where the identification of a mythologised Highlander as the epitome of all that was most noble in the Scottish character had in consequence to wait until the early nineteenth century (Connolly 1995).

The dramatic flowering of patriot and radical politics in the last quarter of the eighteenth century did not involve only the former English of Ireland. The Presbyterians of Ulster had retained something of their cultural distinctiveness, in speech and lifestyle, and also their strong links with Scotland, most notably in the continued attendance of both clergy and laity at Scottish universities. Above all, they had retained their religious distinctiveness. Staunch Presbyterian support for William III had helped to ensure that, from the 1690s, direct religious harassment was no longer permitted. But Presbyterians continued to resent the sacramental test (which excluded from central and local government all those not prepared to take communion in the established church), as well as the continued obligation to pay tithes for the support of the Church of Ireland.

To these grievances against the Anglican establishment were added other features that made Ulster Presbyterian culture a natural base for the development of radical politics. High levels of commercial development in east and central Ulster created an assertive middle class of entrepreneurs, artisans and affluent farmers. Relative prosperity brought good communications and high levels of literacy, facilitating the growth of popular politics. Presbyterian church government was participatory and egalitarian. All these points help explain why Ulster Presbyterians were consistently among the strongest supporters of successive agitations for legislative independence and parliamentary reform. Individual Presbyterians were also active in the newly fashionable rediscovery of traditional Gaelic culture. A people who only a generation earlier had still thought of themselves as the Scots of Ireland were now equipping themselves with their own version of an Irish identity (Stewart 1986).

Meanwhile, Irish Catholics were also, if somewhat less dramatically, redefining their self-image and allegiances. For half a century or more after the fall of James II, the great majority, of all social classes and including both clergy and laity, had continued to support the claims of the exiled Stuart dynasty. After the failure of the British Jacobite insurrection of 1745, hopes of a Stuart restoration were recognised by all but the most zealous to be unrealistic. From the 1750s there were clear signs that the Catholic propertied classes were prepared to come to terms with the existing political order. Commencing in 1759, a succession of Catholic committees campaigned, initially almost timorously, later with more insistence, for a relaxation of the penal laws, while at the same time offering unambiguous declarations of loyalty to the ruling Hanoverian dynasty. They were encouraged in their efforts by growing signs that British government was no longer committed to the maintenance of unqualified Protestant supremacy in Ireland. This was partly because of the decline of anti-Catholicism among the British élite. It also reflected a growing interest in clearing the way for more extensive military recruitment in Ireland, at a time when a succession of large-scale wars was stretching Britain's manpower resources to the limit. There was also at least the possibility that a deferential Catholic population, wholly dependent on the favour of London, might be of use in checking the aspirations of Protestant patriot troublemakers (Bartlett 1992).

The rejection of Jacobitism by the Catholic propertied classes of the mid-eighteenth century was not necessarily matched lower down the social scale. Among the common people notions of subjection to a usurping and oppressive regime, along with hopes of deliverance through the restoration of the rightful dynasty, were perpetuated in poems and ballads. As time went on, other forms of social and political discontent appeared side by side with, or grafted on to, this traditional disaffection. From the 1760s, rural Ireland was disturbed by a series of agrarian protest movements which expressed the tensions created by rising population and the commercialisation of agriculture, and which also reflected the new potential, in a more economically developed society, for popular mobilisation. Initially these agrarian movements were conservative in outlook, seeking to preserve an existing way of life against unwelcome economic innovations. Over time, however, rising living standards, greater mobility, increased literacy and the wider dissemination of the English language opened up the Catholic lower classes to new political ideas, originating from domestic radicalism, republican America and, later, revolutionary France. Popular political awareness was further heightened by the growth of sectarian tension, as Catholic assertiveness and the ambivalence of British government provoked a strong Protestant back-lash. By the early 1790s, the Defenders, originally a lower-class Catholic secret society in south Ulster, had spread across most of the northern half of the country, bringing with them a crude revolutionary ideology in which anti-Protestantism was combined with aspirations to a vaguely imagined overthrow of the social and political order (Smyth 1992: 10–51; Whelan 1996a: 37–42).

The place of Jacobitism in all of this provides a particularly vivid example of the way in which the apparent continuities of Irish political history can conceal what are in fact striking changes in content and definition. Loyalty to the exiled House of Stuart, and the repudiation it necessarily involved of the ruling Hanoverian dynasty, is all too easily assimilated to the image of a long-standing tradition of 'Irish' (and Catholic) resistance to 'English' rule. And indeed it may well be that Irish Jacobitism, like its Scottish counter-part, was among other things a vehicle for a strongly felt resentment at political subordination to those who were perceived as foreigners. But the fact remains that Jacobitism, concerned to set a Scottish dynasty on the united thrones of Great Britain and Ireland, was by definition a British political ideology. Like the earlier service of Irish Catholics in the armies of James II, it derived its whole rationale from an assumption that the three kingdoms of the British Isles would remain under one sovereign. More important still, Jacobitism was inherently conservative. Its political theory rejected Whig notions of a contract between rulers and ruled in favour of the claims of heredity and divine right; in its specifically Irish manifestation it looked back to the aristocratic and fiercely anti-egalitarian world of the pre-plantation Gaelic past. In so far as it contributed to the sort of disaffection represented by the Defenders, this required a radical redefinition in which a dynastic and aristocratic ideology, rooted in the Europe of the *ancien régime*, was recast in terms of the egalitarian republicanism of the late eighteenth-century Atlantic revolutions. Resentment of past wrongs shifted from the overthrow of the Gaelic aristocracy to an imagined dispossession of the Irish people as a whole; deliverance came to mean, not the return of a Stuart monarchy, but the establishment of an Irish republic. The reappearance of expectations of French military assistance masked the transition from the Catholic absolutism of Louis XIV and his successors to the revolutionary republic. In all these respects, the tunes being played by late eighteenth-century opponents of the established order were superficially familiar; but the words being sung to them were wholly new.

CRISIS: THE 'CIVIL WAR' OF 1798

These changes in the political self-definition of Anglican, Presbyterian and Catholic took place against a background of tension and upheaval. The American and French revolutions released new political aspirations while at

the same time making establishments less secure and in consequence potentially repressive. Economic development and population growth intensified
social conflict. The Catholic question was once more the subject of bitter
debate. Patriot successes in 1779–82 had left the whole framework of
Anglo-Irish relations dangerously uncertain. By the 1790s, events were
moving towards a crisis in which all sections of Irish society had to face
fundamental choices. Middle-class radicals had to resolve whether they were
prepared to carry their opposition to a corrupt, landlord-dominated political
system to the point of mobilising the forces of popular discontent, with the
attendant risk of unleashing a social revolution. Protestant patriots had to
decide whether their commitment to a sense of Irish liberties was strong
enough for them to forsake the protection of Great Britain for alliance with a
Catholic majority whose ultimate intentions remained to be discovered.
Presbyterians had to ask whether they were victims along with the Catholics
of an oppressive Anglican establishment, or part of a vulnerable Protestant
minority. Catholics had to decide whether their interests were best served by
joining with Protestant radicals in an attack on the whole political and
social order, or by seeking to outbid their opponents in terms of loyalty to
the crown and the London government. Faced with such questions, persons
of identical religious, cultural and social background gave radically different
answers, producing a complex and rapidly changing pattern of commitment
and uncertainty.

Matters came to a head in the summer of 1798. The insurrection that
took place was subsequently to be assimilated into nationalist political
memory as one of a series of revolts against 'English rule'. In fact it was
above all an Irish civil war. Its aim was the overthrow, not just of the Anglo-
Irish connection, but of the Protestant governing class whose power was
channelled through the very same 'independent Irish parliament' that was to
be the model for subsequent nationalist movements from O'Connell to
Redmond. The insurgent forces included both Catholics and Protestants, the
latter mainly – but by no means exclusively – drawn from the Presbyterians
of the north-east. The army which defeated them consisted of English
regular soldiers and a much larger body of locally raised forces, including the
predominantly Catholic militia.

Behind this national conflict of loyalist and rebel, confusing enough in
itself, it is important to note the intricate patchwork of local and sectional
loyalties. The heartland of the United Irish movement was the two east
coast cities of Belfast and Dublin, with their hinterlands. In Munster,
Connacht, west Ulster, and even parts of Leinster, French-inspired radicalism made little impact. In County Mayo the members of the French
expeditionary force that arrived belatedly at the end of August found a
population who welcomed them, not as standard-bearers of the secular
republic but as soldiers of a traditional Catholic ally, come to liberate
Ireland for Christ and the Blessed Virgin. Even closer to the two radical

capitals, political responses varied greatly. Recent studies of County Wexford have shown how French-inspired republicanism interacted with a range of local political, sectarian, ethnic and factional conflicts (Cullen 1981: 210–33; Whelan 1996b). In the north-east, radicalism took on very different meanings in the context of the two main strands of Presbyterian doctrine. For 'New Light' Presbyterians, commitment to equality and political reform went hand in hand with a rational Enlightenment-influenced theology. But the United Irish movement also attracted 'Old Light' Presbyterians, committed to a traditional theology in which the overthrow of landed magnates, placemen and the Anglican establishment was to be interpreted in terms of the defeat of Antichrist and the triumph of a political order based on the Old Testament covenant between God and man (Miller 1978).

The traumatic events of 1798 had a dramatic impact on the pattern of political self-definition and alignment. Eighteenth-century patriotism had been born out of a confidence that a self-governing Ireland would be one controlled by the Protestant propertied classes, whose economic and cultural advantages made them the natural rulers of Irish society. Already during the 1790s, however, the rapid politicisation of the Catholic masses had made this assumption increasingly doubtful. The naked sectarian hostility displayed in parts of the south during the insurrection gave new substance to traditional fears of a vengeful and politically revived Catholic majority. The new legacy of grievance, built up on both sides by bloodshed and intimidation, also severely limited the scope for further projects for political co-operation between Catholic and Protestant. In the immediate aftermath of the rebellion, patriot sentiment was still strong enough for a substantial section of Protestant opinion to oppose the Act of Union. The Catholic church authorities and the majority of propertied Catholics, by contrast, welcomed a measure that liberated them from the authority of a parliament dominated by reactionary Irish Protestantism, and which they had been led to believe might be followed by full Catholic emancipation. Thirty years later, these alignments had been almost wholly reversed. Daniel O'Connell's campaign for repeal of the Act of Union drew its support from Catholics alienated by the failure of British government to deliver the expected progress towards religious equality, as well as by economic and social grievances which they had been persuaded to identify with the cause of Irish self-government. On the other hand, all but a small minority of Protestants had come to see in the union their only protection against an increasingly assertive and well-organised Catholicism.

AFTER THE UNION: AMBIGUOUS IDENTITIES

It would be wrong to see the changes in political outlook that followed 1798 as clearing the ground for the emergence of a polar opposition between Catholic nationalism and Protestant unionism. It is possible, in the first place, to exaggerate the impact of the rebellion. The scale of violence, and the bitterness created, was undoubtedly considerable. From another point of view, though, what is remarkable is how little immediate change there was in the pattern of Irish politics. For more than two decades following the Act of Union, parliamentary elections were the almost exclusive preserve of the same landed families that had dominated the old Irish parliament, deploying the votes of a compliant tenantry, Catholic as well as Protestant, in socially exclusive and largely non-ideological contests. Thereafter, from the mid-1820s to the mid-1840s, O'Connell's campaigns, first for Catholic emancipation and then for repeal of the union, introduced a strikingly new style of popular political agitation. Yet this development never affected more than a proportion of Irish constituencies: O'Connell, at his peak, led forty-two MPs out of a total Irish representation of 105. After his death and the collapse of repeal, moreover, the 1850s and 1860s saw a marked resurgence in the political influence of localism, personality and proprietorial interest. Even the Home Rule party, whose triumph in the general election of 1874 seems at first sight to have marked the end of this golden age of traditional politics, turns out on closer inspection to have contained many members of the landed élite, opportunistically enlisted under a new flag. Indeed, it was not until 1886 that Irish constituencies sent to Westminster a selection of MPs not dominated by the Protestant landowning and professional classes. The resilience of this group testifies to the extent to which conflicts of political and religious identity continued to exist side by side with a web of personal and local loyalties, ties of clientage and deferential acceptance of the wishes of social superiors (Hoppen 1984).

If the decline in the authority of the traditional landed élite should not be predated, neither should the polarisation of national allegiances between Catholic and Protestant. The terrifying experiences of 1798 may have spelled the end for a confident and broadly based patriot tradition, but Protestant political nationalism persisted in the Young Ireland movement of the 1840s. Later still, there was the extensive Protestant support attracted by the early Home Rule movement, at a time when the disestablishment of the Church of Ireland had led many of its adherents to question the benefits of the union. Protestant–Catholic co-operation on issues other than self-government also continued, notably in the Tenant League of the early 1850s. In addition, there was a broader tradition of what R. F. Foster (1993: 62) has called 'hard headed' landlordism: a pragmatic willingness among sections of the Protestant élite to support progressive measures, including even reform of the law of landlord and tenant and schemes for limited self-government, as a means of preserving their social position and something of their political authority.

Protestant political nationalism may have been a minority and declining movement, but Protestant literary and cultural nationalism was, however, a central element in nineteenth-century intellectual life. Prior to the formation and rise to prominence at the very end of the century of the GAA and the Gaelic League, Catholic nationalism was characterised by a pragmatic acceptance of the fact of Anglicisation: O'Connell's repealers, the Fenians, Home Rulers under Parnell and after, all focused their attention on the attainment of a purely political autonomy, supported by arguments based on history, legal precedent, justice and practicality. By contrast, it was the Protestant patriots of the late eighteenth century who had first linked the assertion of Irish constitutional rights to an exalted vision of the Gaelic past. The Young Irelanders further developed the notion of an Irish claim to independence, based on cultural identity as well as geography and legal title, and drew on the Gaelic literary heritage as raw material for political symbolism and propaganda. In other cases, the rediscovery and popularisation of that heritage was the work of Protestant writers such as Samuel Ferguson and Standish O'Grady, who combined strong unionist politics with a deep imaginative involvement in the world of Gaelic history and myth.

Recent studies have stressed the extent to which this Protestant cultural nationalism should be seen as a strategy for self-defence. The Young Irelanders sought to emphasise a cultural definition of nationality as an antidote to the growing sectarianism of Irish public life. More than this, Thomas Davis's attempt to rekindle in his co-religionists an interest in Gaelic culture and an enthusiasm for national rights can be read as an appeal to the Protestant propertied classes to resume the political leadership of a patriotic public opinion that they had allowed to slip by default into the hands of the Catholic middle classes and the Catholic clergy. With Ferguson and O'Grady, the political message is stated more openly: the Gaelic Ireland of their imagination, a stable, hierarchical society in which lord and peasant were bound together by shared cultural values and ties of mutual respect, was an ideal to be set against the contemporary reality of an unruly democratic politics and an upstart Catholic bourgeoisie. This appropriation of the Gaelic past as an image of élite hegemony, still evident in the political poetry of Yeats, is a further illustration of the way in which 'tradition' was repeatedly reshaped in the service of current political needs (Cairns and Richards 1988: 25–41, 97–103).

Any realistic analysis of nineteenth-century Irish political life must also take account of the extent to which this was dominated by a system of allegiance and identification derived from the British politics of the day. For most of the century following the Act of Union, the majority of Irish political representatives identified themselves in terms of the two main United Kingdom political parties, Whig and Tory, later Liberal and Conservative. Support for both was based on a complex of interlocking influences. The Conservatives were to a substantial extent the Protestant party, and more particularly that of the established church, but they also attracted the votes of all those who, for whatever reason, wished to uphold the entrenched social and political order, as well as of those susceptible to proprietorial influence or the claims of paternalism. Liberals were traditionally more sympathetic to Catholic interests and, as such, significantly more likely to attract Catholic votes where these were freely cast. But they were also the party of the reform-minded middle classes and artisanate, of the tenant rather than the landlord, and of Protestant dissenters rather than the established church. There were also issues on which Conservatives were closer than Liberals to reflecting Catholic views. In 1859, for example, British Liberal support for the Italian nationalists whose military successes threatened the Papal States led Catholic voters to give the Conservatives their best-ever Irish result. Against this fluid but nevertheless enduring pattern of British party politics, attempts to promote a distinctively Irish pattern of political allegiances made only limited headway. O'Connell's repeal movement of the 1830s and 1840s moved uneasily back and forth between being a wholly separate party and becoming part of the broad Whig–radical coalition, while the attempt to create an Independent Irish Party in the 1850s foundered as members drifted into support of a Liberal government.

MANIPULATIONS OF IDENTITY: LINEAR NARRATIVES

This background makes it possible to appreciate the scale of the political transformation accomplished by the rise of the Home Rule movement. By the mid-1880s, Irish politics had been reshaped round a simple dichotomy in which religious and political loyalties could be taken as largely inter-changeable. On one side there was a powerful Nationalist party, mobilising the votes of the overwhelming majority of Catholics and controlling every parliamentary seat outside Ulster, Trinity College and, occasionally, the affluent suburbs of south Dublin. On the other was a Unionist party, closely allied to, but distinct from, the British Conservatives, mobilising the great majority of Protestants in defence of the union. The suddenness of the change, and the finality with which other political options were closed off, was particularly apparent in Ulster. In the 1870s and early 1880s, the Ulster Liberals had enjoyed unprecedented electoral success. In 1874 they had captured six out of twenty-nine Ulster seats, in 1880 nine, in each case drawing on a combination of Catholic and Protestant votes, and capitalising on the potential of rising agrarian discontent to alienate Protestant farmers from the landlord-dominated Conservative party. In 1885, however, the Liberals disappeared from the electoral scene in Ulster as elsewhere, as Catholic supporters deserted to Home Rule and Protestants to unionism (Walker 1989).

The immediate means by which this transformation was effected are well known. Isaac Butt had already in the 1870s skilfully won the support of militant nationalists – the Fenians – for his parliamentary Home Rule movement; nevertheless, the party he led was still one in which the traditional parliamentary classes were well represented: no less than forty-two of the sixty Home Rulers returned in 1874 were landowners or professional men. The real change came after 1879 when Parnell successfully linked the cause of Home Rule to the major campaign of agrarian protest taking shape as farmers felt the effects of the worst agrarian depression since the Famine. By placing himself at the head of the tenant movement, Parnell propelled himself from the extremist fringes of the Home Rule movement to its leadership, while at the same time giving parliamentary nationalism a mass following on a scale, and of a solidity, never previously achieved. The changed character of the party was evident in the social backgrounds of the eighty-six Home Rule MPs elected in 1885: only five were landowners, as compared to twenty-three in 1874, while forty-one were drawn from the lower professions or from the shopkeeper/farmer class.

The spectacular success of nationalism in supplanting other alignments, across little more than a decade, owed much to Parnell's political skills. The opportunistic exploitation of the land agitation reflected his ability to combine the politics of the possible with a militant rhetoric in a way that secured him the support of a broad spectrum of opinion, from Catholic bishops and the comfortable middle classes to Fenians and agrarian radicals. But his achievements were made feasible by broader changes in attitudes and ideas. The second half of the nineteenth century saw the development and popularisation of nationalist historical writing, in which the web of changing identities with which we have been concerned was recast as a linear narrative of Irish resistance to English rule. More specifically, the linking of the issues of land and home rule depended on the perfection of a coherent mythology, already beginning to take shape in the Defenderism of almost a century earlier, in which the sixteenth- and seventeenth-century dispossession of the Gaelic and Old English élites was reinterpreted as the dispossession of the Irish people as a whole. This legitimised the claims of the tenant farmer while undermining those of the landlord. The mythology of the Land War of the early 1880s also encouraged a new sense of collective identity. Large and small farmers, the landless and the land-poor, as well as urban groups to whom the farmers' problems were of no direct concern, were taught to see themselves as united in a joint struggle for lost ancestral rights. In short, the linking of land and home rule created an imaginary community, possessed of a strong sense of collective identity based on historic wrongs and current grievances (Anderson 1991).

The other major element in what is today considered the nationalist tradition, an identification with the Gaelic past, was added only later. The GAA, meeting a real social need in a more affluent countryside whose traditional games and pastimes had been largely lost, rapidly grew into a genuinely popular organisation. The literary and linguistic movement, however, initially appealed to a relatively small circle; it was only later, with the triumph during and after World War I of what had up to then been the minority ideology of separatist republicanism, that it entered the mainstream of Irish nationalism. Before this development, cultural revivalism had relied for much of its support on two specific groups. One, continuing a tradition going back to Ferguson and the Young Irelanders, consisted of middle- and upper-class Protestants anxious to reaffirm their place in Irish society at a time when they had been politically marginalised, and when an increasingly strident political rhetoric identified Irishness exclusively with Catholicism. The other comprised urban, white-collar workers, often themselves risen from traditional rural backgrounds through the newly developed system of mass education and public exami-nations. These had been the main beneficiaries of the rapid growth in preceding decades of an Anglicised and commercialised Ireland, but they also comprised the group most affected by the accompanying sense of alienation and loss of cultural roots (Waters 1977).

To both these groups, suffering from their separate crises of identity, cultural nationalism offered the possibility of a willed identification with the Irish-speaking peasantry of the impoverished far west, one that tran-scended the realities of class, lifestyle, economic environment and – in some cases – religion and ethnic origin. What was involved, inevitably, was not cultural revival, but rather reinvention. The Gaelic games propagated by the GAA did not mark a return to the rough-and-tumble participatory sports of the Irish countryside; instead they represented a spectator sport adapted to the needs of an emerging consumer society. The literary culture celebrated by the Gaelic League was likewise a bowdlerised version of authentic tradition, adapted to conform to the expectations of middle-class Victorian morality (Cullen 1981: 255–6; Hutchinson 1987: 114–87; Garvin 1987: 78–106).

The rapid progress of nationalism in the 1870s and 1880s thus depended on the creation, by the manipulation of historical myth and political symbol, of a new collective identity. The same was true of the other side of the political dichotomy that had now overridden all other identifications. Unionism, no less than nationalism, depended on an inte-gration of groups that had previously had their own separate allegiances and history. It brought together Anglican and Presbyterian, landlord and tenant, urban workers and middle-class business interests, all united by a shared allegiance to Protestantism and the maintenance of the union. No part of this network of alliances could be taken for granted. Presbyterians retained both a strong sense of their separate religious identity and a capacity for resentment at what they still saw as an Anglican-dominated establishment. In the late 1890s and early 1900s, with the threat of Home Rule temporarily in abeyance, there were breakaway movements and electoral revolts by tenant farmers and urban workers alienated by a leadership too concerned with the interests of landlords and employers (Jackson 1987; Gray 1985). The development from 1910 of a new Home Rule crisis produced a rapid closing of ranks, and the violence of the years 1919–23 reinforced this restored cohesion. But urban working-class dissatisfaction was to reappear at intervals, most notably in the combined action of Catholic and Protestant unemployed in the outdoor relief riots of 1934, and in electoral successes by the Northern Ireland Labour Party in the post-war period. The practical significance of these episodes, despite the attention they have received from nationalist and socialist writers alert for any sign of a crack in the monolith of Ulster unionism, should not be exaggerated: hopes that the politics of class could replace those of religion and national allegiance repeatedly proved illusory. Their importance is rather as reminders of the success of the unionist class alliance in erasing alternative lines of cleavage and identification.

As with nationalism, so the construction from the 1880s of this new unionist collective identity was assisted by the exploitation of both real and invented history. It was at this point that the Orange Order, which in the middle decades of the century had become almost entirely a plebeian movement, regained élite leadership and became once again a vehicle of vertical integration. Its rituals, ballads and parades – centred on resistance to the threat of popish tyranny under James II but also involving memories of the 1640s and of 1798 – provided a vision of Irish Protestant, and more specifically Ulster Protestant, history in terms of repeated episodes of providential delivery and heroic self-defence. At the same time, this superficial similarity should not be taken at face value. Unionism in the late nineteenth century and after may have drawn on a range of symbols from the past, but the political and cultural identity that these were intended to support remained ambiguous. Early unionist propaganda wavered uncertainly between claims to Britishness and expressions of a distinctively Irish loyalism (Jackson 1989: 8–16). Later, following what had initially been a pragmatic acceptance of the facts of religious geography, there were to be attempts to promote the idea of a separate Ulster identity, similar uncertainties of direction remaining evident in the present day.

CONCLUSION

A survey of political identifications and allegiances across the centuries before partition thus offers little support to the notion of two coherent historical and cultural traditions locked in long-standing conflict. In the first place, the implied symmetry of the conflict is misleading. As explored further in Chapter 10, unionists, uncertainly balanced between self-assertion and dependence on Britain and seeking to defend their right to self-determination against opponents who had appropriated the vocabulary of nationalism to themselves, have failed to match the coherent fusion of real and imagined elements from cultural and political history achieved by late nineteenth-century Irish nationalism. Instead they remain condemned to be clear on what they oppose, far less so on what they stand for. Yet the nationalist 'tradition' too remains, for all its superior coherence, the outcome of a specific process of invention and reinterpretation. In the space of a few decades, historical memory was reshaped to obscure older allegiances and identifications, redefine sectional grievances as national wrongs, and recast complex past conflicts in the mould of the present. New definitions were created of what constituted Irishness, and imaginative links were established with a sanitised and idealised version of a way of life whose disappearance had been part of the very process of modernisation that had made nationalism itself possible.

To say this is not to suggest that the 'two traditions' model of the problems of contemporary Northern Ireland lacks value. That a tradition is wholly or in part invented does not preclude it from becoming a force in its own right. Indeed, it could be argued that such invention is an essential part of the creation of any workable political community. Yet neither should we accept without question the permanent validity and exclusive claims of definitions of identity laid down in specific circumstances in the late nineteenth century. Nor should we surrender to the facile cliché that the modern Irish are prisoners of an inflexible past. What is demonstrated by the kaleidoscope of identities and allegiances examined here is rather the flexibility of tradition, and the ability of successive generations to reshape the past so as to serve the needs of the present.

REFERENCES

Anderson, B. (1991) *Imagined Communities: Reflections on the Origin and Spread of Nationalism*, revised ed., London: Verso.

Bartlett, T. (1992) *The Fall and Rise of the Irish Nation: The Catholic Question 1690–1830*, Dublin: Gill and Macmillan.

Brady, C. (ed.) (1994) *Interpreting Irish History: The Debate on Historical Revisionism*, Dublin: Irish Academic Press.

Cairns, D. and Richards, S. (1988) *Writing Ireland: Colonialism, Nationalism and Culture*, Manchester: Manchester University Press.

Clarke, A. (1978), 'Colonial identity in early seventeenth-century Ireland', in T. W. Moody (ed.) *Nationality and the Pursuit of National Independence*, Belfast: Blackstaff.

Connolly, S. J. (1992) *Religion, Law and Power: The Making of Protestant Ireland 1660–1760*, Oxford: Clarendon.

—— (1995) 'Popular culture: patterns of change and adaptation', in S. J. Connolly, R. A. Houston and R. J. Morris (eds) *Conflict, Identity and Economic Development: Ireland and Scotland 1600–1939*, Preston: Carnegie.

Cosgrove, A. (1979) '"Hiberniores ipsis Hibernis"', in A. Cosgrove and D. McCartney (eds) *Studies in Irish History*, Naas: Leinster Leader.

Cregan, D. (1995) 'The Confederate Catholics of Ireland: the personnel of the Confederation 1642–9', *Irish Historical Studies* 29: 490–509.

Cullen, L. M. (1981) *The Emergence of Modern Ireland 1600–1900*, London: Batsford.

Cunningham, B. (1986a) 'Native culture and political change in Ireland 1580–1640', in C. Brady and R. Gillespie (eds) *Natives and Newcomers*, Dublin: Irish Academic Press.

—— (1986b) 'Seventeenth-century interpretations of the past: the case of Geoffrey Keating', *Irish Historical Studies* 25: 116–28.

Deane, S. (ed.) (1991) *The Field Day Anthology of Irish Writing*, London: Faber.

Ellis, S. G. (1986) 'Nationalist historiography and the English and Gaelic worlds of the late Middle Ages', *Irish Historical Studies* 25: 1–18.

Foster, R. F. (1993) *Paddy and Mr Punch*, London: Allen Lane.

Frame, R. (1993) ' "Les Engleys nees en Irlande": the English political identity in medieval Ireland', *Transactions of the Royal Historical Society* ser. 6, 3: 83–103.

Garvin, T. (1987) *Nationalist Revolutionaries in Ireland 1858–1928*, Oxford: Clarendon.

Gray, J. (1985) *City in Revolt: James Larkin and the Belfast Dock Strike of 1907*, Belfast: Blackstaff.

Hayton, D. H. (1988) 'From barbarian to burlesque: English images of the Irish *c.* 1660–1750', *Irish Economic and Social History* 15: 5–31.

Hoppen, K. T. (1984) *Elections, Politics and Society in Ireland 1832–1885*, Oxford: Clarendon.

Hutchinson, J. (1987) *The Dynamics of Cultural Nationalism: The Gaelic Revival and the Creation of the Irish Nation State*, London: Allen and Unwin.

Jackson, A. (1987) 'Irish unionism and the Russellite threat 1894–1906', *Irish Historical Studies* 25: 376–404.

—— (1989) *The Ulster Party: Irish Unionists in the House of Commons 1884–1911*, Oxford: Clarendon.

Kiberd, D. (1995) *Inventing Ireland: The Literature of the Modern Nation*, London: Cape.

Lambkin, B. K. (1996) *Opposite Religions Still? Interpreting Northern Ireland after the Conflict*, Aldershot: Avebury.

Leerssen, J. Th. (1988) 'Anglo-Irish patriotism and its European context: notes towards a reassessment', *Eighteenth-Century Ireland* 3: 7–24.

Lyons, F. S. L. (1979) *Culture and Anarchy in Ireland 1890–1939*, Oxford: Clarendon.

Miller, D. W. (1978) 'Presbyterianism and "modernization" in Ulster', *Past and Present* 80: 66–90.

Smyth, J. (1992) *The Men of No Property: Irish Radicals and Popular Politics in the Late Eighteenth Century*, London: Macmillan.

—— (1993) ' "Like amphibious animals": Irish Protestants, ancient Britons 1691–1707', *Historical Journal* 36: 785–97.

Stewart, A. T. Q. (1986) 'The harp new strung: nationalism, culture and the United Irishmen', in O. MacDonagh and W. F. Mandle (eds) *Ireland and Irish-Australia*, London: Croom Helm.

Two Traditions Group (1983) *Northern Ireland and the Two Traditions in Ireland*, Belfast: Two Traditions Group.

Vance, N. (1981) 'Celts, Carthaginians and constitutions: Anglo-Irish literary relations 1780–1820', *Irish Historical Studies* 22: 216–38.

Walker, B. (1989) *Ulster Politics: The Formative Years 1868–86*, Belfast: Ulster Historical Foundation.

Waters, M. J. (1977) 'Peasants and emigrants: considerations of the Gaelic League as a social movement', in D. Casey and R. Rhodes (eds) *Views of the Irish Peasantry 1800–1916*, Hamden, Conn.: Archon Books.

Whelan, K. (1996a) *The Tree of Liberty: Radicalism, Catholicism and the Construction of Irish Identity 1760–1830*, Cork: Cork University Press.

—— (1996b) 'Reinterpreting the 1798 Rebellion in County Wexford', in D. Keogh and N. Furlong (eds) *The Mighty Wave: The 1798 Rebellion in Wexford*, Dublin: Four Courts Press.

4

WRITING IRELAND

Literature and art in the representation of Irish place

Patrick J. Duffy

PORTRAIT OF THE ARTIST AS WITNESS

The territory of Ireland, with all its nationalist and all its gothic graves, with all its mouldering estates and emerging farms, its Land Acts and its history of confiscations, was in need of redefinition by the early years of the century.

(Deane 1994: 34)

Cultural geography is a reaction against earlier preoccupations with objectivity and the mechanistic and 'dehumanised' approaches of positivism. Livingstone (1992: 337) suggests that 'the human had been elided in "human" geography . . . replaced by rational economic actors . . . [and] sucked dry of desires, meanings and emotion'. In its concern with psychological and emotional attachments to place and their economic and political manifestations, contemporary cultural geographers question how ' "ordinary" people leading "ordinary" lives encounter, perceive and perhaps reflect upon the spaces, places and environments all around them' (Cloke *et al.* 1991: 81). The 'sense of place' accruing from the ways in which people experience representations of present and past landscapes is a fundamental part of territorial identity and of geographical understanding. Its converse, 'placelessness' (Relph 1976), captures the essential geographical emptiness, perhaps even meaninglessness, of societies – or elements thereof – which lack a unifying narrative of place.

One way in which geographers have sought to elucidate and illuminate place identity and place experience has been through creative art – both literature and painting. Writers and artists are both witnesses to our world but also products of it, possessing qualities of insight which can be mustered in helping to understand the diversity of place and the contested meanings that can be attributed to it. Because they can express sublime emotions like love, hate or fear and are similarly capable of articulating manifestations of place, society rewards and honours its artists, who are seen as interpreters of national culture. The role of the artist as witness and interpreter of place, landscape and identity, can be broadened beyond mere reflection or revelation. To a very significant extent our past and present views of Ireland and Irishness have been shaped by readings of literature and art (Dunne 1987).

Ireland has produced some of the most illustrious writers in the English language over the past two centuries, much of their inspiration provided by the meanings of place and landscape in constructs of Irishness. W. B. Yeats, for example, was fond of quoting Turgenev in defence of his artistic role: 'The cosmopolitan is a nonentity – worse than a nonentity; without nationality is no art, nor truth, nor life, nor anything' (quoted in Brown 1972: 14). Yeats had no illusions about the ideological linkage between art and society. For much of his life he was actively involved in Irish political and cultural life, his work with the Abbey Theatre and the Irish Literary Revival in the early twentieth century, for example, being an expression of his commitment to a public role for drama, poetry and art in the negotiation of Irish identities. Again, for Yeats, the 1916 Rising was an event inspired and led by poets and therefore a demonstration of the importance of art in public life.

James Joyce also saw himself and his writing as occupying a significant role in Irish life. At the end of *A Portrait of the Artist as a Young Man*, Stephen assumes the role of presiding genius: 'I go to encounter for the millionth time the reality of experience and to forge in the smithy of my soul the uncreated conscience of my race' (1993a: 191). Joyce was the pure artist, whose duty in the search for truth was to stand alone, untrammelled by societal conventions: 'Until he has freed himself from the mean influences around him – sodden enthusiasm and clever insinuation and every flattering influence of vanity and ambition – no man is an artist at all' (cited in Mason and Ellmann 1959: 69). It was, perhaps above all else, his disenchantment with the Gaelic Revival and the escalating censorship of the 'rabblement' of Dublin and Ireland that forced Joyce to abandon Ireland for good.

Among contemporary writers, Seamus Heaney both sees himself and is seen as an important articulator of Irish consciousness. In interrogating the role of the poet, he raises important issues about his situation in Ireland and his relationship 'to his own voice, his own place' among the confusions of Ulster. 'One half of one's sensibility is in a cast of mind that comes from belonging to a place, an ancestry, a history, a culture' (1980: 13). While Heaney would see himself as an Irish poet, in Northern Ireland he has been characterised as 'a British subject living in Ulster' (Heaney 1990: 23).

As these examples among Ireland's foremost writers suggest, Irish place and landscape have been variously constructed and interpreted to fulfil the changing requirements of particular segments of society, both inside and outside the island. In this respect, literary texts can be regarded as signifying practices, which interact with social, economic and political institutions so that they 'are read, not passively, but, as it were, rewritten as they are read'. Nor are they merely mirrors to a 'reality outside themselves' but communicate and produce meanings which vary through time and across and even within cultures (Barnes and Duncan 1992: 5–6). Sheerin (1994) argues that places and landscapes are narrative constructions produced by writers and often more real than reality itself, so powerful and influential is the role of the artist. Without Yeats, for example, Inishfree would be a nameless place. He is the supreme example of an artist setting out to construct a deliberate, symbolic landscape allegory of identity, impressing himself on a landscape like a 'phase of history' (O'Connor 1950: 256).

Thus, as with its geography and history, the writing of Ireland and its landscapes is also implicated in the flexibilities and fluidities of contested constructions of Irish identity. Apart from generic-type dichotomies such as rural and urban, east and west, there are the much more problematic cultural and political confusions symbolised by the plethora of qualified Irish – Anglo-Irish, Protestant Irish, Catholic Irish, Gaelic Irish, 'West Brit', Ulster Irish, Ulster Scots, Scots Irish, Northern Irish. In writing of the 'doubleness of our focus' in Ireland and 'our capacity to live in two places at the same time and in two times at the one place', Heaney (1990: 22) underscores this ambiguity. Again, Foster (1989: 12) points to the social and cultural diversity of nineteenth-century Ireland, subsumed by Celtic revivalism in the late nineteenth century, itself a powerful example of a nationalist trope of identity. He suggests that the works of John Banim, Gerald Griffin, Anthony Trollope, George Moore and Somerville and Ross, among others, all reflect on a complex and varied world that has long been lost to our sight. The remainder of this chapter focuses upon a variety of themes relating to the diverse meanings of Irish identity and landscape as represented in the works of artists. These range through the phases of eighteenth- and nineteenth-century Romanticism, the contrived Celtic twilightism of the West, notions of rural arcadia – both peasant and big house – contested representations of the North, and emigrant landscapes.

NINETEENTH-CENTURY ROMANTICISM AND THE MYTH OF THE WEST

Nineteenth-century artistic representations of Ireland and its landscapes reflected the country's subservient economic and political relationship with Britain, patronage of the arts being largely dictated by the preferences and priorities of Victorian England. Often shaped by the tenets of the Romantic movement, which was strongly English in origin and development, nine-teenth-century images of Ireland are important in discussing the emergence of Irish identity because of their influence on the iconography of twentieth-century nationalism. The Gaelic Revival, the 'West of Ireland' imagery of Synge and Yeats, and even the Catholic Motherland visualised by Éamon de Valera, owe much to the ethos of Romantic mysticism and exoticism engendered by some nineteenth-century artists and writers. Their representations of Ireland as exotic, sublime and picturesque reflect the way in which artistic imagery incorporated 'blindnesses and silences', dictated by the market and English sensibilities (Duffy 1994). The poverty and squalor of many Irish landscapes, often the most picturesque and romantic marginal and mountain wastelands, was a problem in such renditions. Artists like George Barret avoided these places or painted out the squalor, while landowners commissioned views which avoided the ugliness of destitution. Writers such as William Carleton (1865) also tried to make peasant poverty acceptable through romantic landscape settings, or by incorporating scenes of lively melodrama into their work.

English perceptions of the Irish were largely based on purely Romantic constructions of Otherness. Celtic cultural distinctiveness was reflected in the wildness and strangeness of an exotic, imaginative race, inhabiting untamed landscapes of 'horrible beauty', a people very different from the practical and pragmatic English. Such traits imbue much nineteenth-century literature, beginning with Maria Edgeworth's pioneering *Castle Rackrent*, published in 1800, in which Ireland is represented – or misrepresented – as a rather surreal place. William Steuart Trench's memoirs of life as a land agent – *Realities of Irish Life* (a title, like Carleton's *Traits and Stories of the Irish Peasantry*, which is loaded with iconic significance) – recalled life 'surrounded by a kind of poetic turbulence and almost romantic violence', which he suspected probably sounded incredible to an English audience (Trench 1868: vii). In a real sense, the single-minded modernism of James Joyce ran headlong into this residual Victorianism in the early years of the twentieth century. He failed to have *Dubliners* published in Ireland because the publishers wished to excise passages which might offend the Dublin or English bourgeois readership. Ultimately, Joyce's refusal to compromise his artistic principles and his commitment to his own representation of the realities of Dublin life, was to mark an important break in ways of writing about Ireland.

The image and – ultimately – myth of the West was a central motif in the Irish cultural nationalism which evolved towards the end of the nineteenth century. The West was represented as containing the soul of Ireland – in Yeats's construction, a fairyland of mist, magic and legend, a repository of Celtic consciousness. J. M. Synge's plays (1910), if much execrated by emerging Dublin bourgeois Catholic nationalism, extolled the virtues of a primitive society on the edge of Europe, punctuated by violence and lawlessness, its landscapes of savage mountains inhabited by wild men and promiscuous women. This literary narrative of the 'wild West' was complemented by the work of artists like Paul Henry, who observed of early twentieth-century Achill: 'the habits and ways of this remote community surrounded by savage rocks and treacherous seas, provided me with all I required as a painter' (Cosgrove 1995: 101). His desolate landscapes of thatched houses and blue mountains became part of the nationalist iconography of the Free State. (Interestingly, Henry was also commissioned to paint representative landscapes for the new Northern Ireland tourist authority.) Such representations of the West were further elaborated in the era of film: Robert O'Flaherty's *Man of Aran* (1934) is a cinematic reflection of the theme of wild beauty in the western isle. In *The Quiet Man* (1952), John Ford celebrates the West as a passionate, patriarchal, violent society, while David Lean's *Ryan's Daughter* (1970) and Jim Sheridan's *The Field* (1992) can be seen as more recent manifestations of this self-same mythology.

The West has continued, therefore, to carry burdens of authenticity into the twentieth century, the 'bearer of the authentic, quintessential Irish identity, encoded in a landscape different to the industrialised, modernised landscapes of contemporary Britain' (Whelan 1993: 42), the source of Yeats's 'filthy modern tide'. For de Valera, the imagery of the rural Irish-speaking West defined the essence of Irish nationhood, while Padraic Pearse had a cottage in Rosmuc in Connemara where he learnt Irish and communed with the western landscape. His short stories and some of his poems reflect the pastoral simplicity that inform such perceptions of the West.

> Old Mathias heard the roar of the waves on the rocks, and the murmur of the stream flowing down and over the stones. He heard the screech of the heron-crane from the high rocky shore and the lowing of the cows from the pasture, and the bright laughter of the children from the green . . . [and] the clear sound of the bell for the Mass that was coming to him on the wind in the morning stillness.
>
> (Pearse 1924: 229)

The artist Seán Keating complemented the literary Wests of Pearse and de Valera, his paintings portraying an heroic society inhabiting its wild landscapes. His *Men of the West*, for example, depicting republican gunmen in Connemara, suggests parallels between the 'Wild West' in America and in Ireland – places where, in artistic cenceptualisations, national identity was forged (Gibbons 1996: 23).

Thus, as Whelan (1993: 40) suggests, the West today reflects a clash of insider and outsider views: it has been burdened with many-layered representations both of outsiders like Synge and Henry with their constricted urban backgrounds, and of Pearse and de Valera with their preconceived notions of Irishness. Leerssen (1994: 9) uses the term, 'chronotype', to conceptualise the myth of the West, observing that it is a characteristic of European civilisation that peripheries are represented as beyond time and space. Thus the West of Ireland became a Celtic fringe, 'a transitional zone between the historical reality of the mainland [east of the Shannon] and the eternal dreamscape of the ocean'. A report in *The Irish Times* (1 November 1994) on urbanites moving to the West voiced exactly these sentiments: 'There's something about the Shannon – when you cross that river you leave Europe. There's a softness, a gentleness, a civilisation here.' While this representation of the West has been very much a carefully crafted dream-world, one which has borne the brunt of national myth for more than a century, the inside view is rather different. Contrasting with the romantic idealisations of the chronotype was the reality of emigration – as the myth of the West was being constructed, its population was leaving (Duffy 1995). In company with many other parts of the West, the inhabitants of Henry's romantic Achill have long-established migrant links with Britain.

RURAL IRELAND

Overarching the idealisation of the West in artistic representations of Ireland is what might be called the myth of the rural, a narrative which has echoes of a more universal allegory of the communality and pastoral tranquillity of the rural idyll in the face of ever-expanding urbanisation. The preoccupation with rural imagery in Ireland can be traced back to nineteenth-century searches for an identity as Other to English industrial urbanism. Certainly the iconography bore some resemblance to reality in that Ireland was a rural periphery, exporting food products and people to the urban cores in England, Scotland and the New World. However, the myth of the rural had largely élitist origins. It was principally Yeats's literary movement which glorified the rural aesthetic as the authentic source of Irishness: the 'Lake Isle of Inishfree' was to be his rural retreat from the 'pavements grey' of the city. The dominance of rural themes in Irish literature has continued throughout the twentieth century, consolidating the stereotypical representation of the real Ireland as a rural place while reflecting, for a while, the demographic reality of a southern society in which the urban proportion of the population remained relatively small. Some of these images fed into a kind of rural nostalgia which survives to the present day (Taylor 1988). For example, Mary Carbery, author of *The Farm by Lough Gur*, came from a well-to-do County Limerick agricultural family. Although her work has a cultural authenticity derived from a strong community value-system, it is also deeply imbued with the romanticism of rurality. The book carries with it 'all the mists and memories, all the scents and stings of the Irish countryside . . . [which] will reveal . . . to the English reader what England might have been before the fairies were expelled and the parsons ceased from conjuring', said Shane Leslie in a somewhat Yeatsian introduction (Carbery 1938: xv).

Many writers, however, have provided less than idyllic interpretations of rural life. Frank O'Connor, Sean Ó Faoláin, Patrick Kavanagh and Sam Hanna Bell are among those who have written about the contested nature of rural society in which ownership and possession of land were the dominant themes. The land legislation of the late nineteenth and early twentieth centuries had conferred ownership on the occupiers of the farms and much of rural life was subsequently shaped by the dictates and hardships of farm life.

In the novel, *Tarry Flynn* (1948), and many of his poems, Kavanagh, for example, depicts the tyranny of landownership in the lives of the people in the 1930s and 1940s. Maguire, the protagonist of his poem *The Great Hunger* (1964: 34–59), reflects on the social and emotional wasteland that imprisonment in his farm has brought him, as well as the irony of the iconic urban myth of the peasant 'in his little lyrical fields'. John B. Keane's play *The Field* (1966) is a similar evocation of the rural West as a violent place 'beyond the pale'. The pessimistic images of writers like Kavanagh contrasted with official attitudes to the land, encapsulated in de Valera's and Fianna Fáil's ideological glorification of the small farm which was embodied in the 1937 Constitution. The early 1940s (when *The Great Hunger* was first published and then banned by the Censorship Board) saw some extreme statements of this rural small-farm ideology, allied with right-wing Catholicism. These culminated in de Valera's famous imagery of 'a land whose countryside would be bright with cosy homesteads' (quoted in Brown 1981: 146). Such narratives were very much at odds with Kavanagh's world; 'sad, grey, twisted, blind, just awful' (1971: 15). His poem 'Stony grey soil' condemns both rural landscape and society: 'O stony grey soil of Monaghan,/The laugh from my love you thieved . . . /You flung a ditch on my vision/Of beauty, love and truth' (1964: 82–3).

At the same time, rural landscape and community were also represented in overtly lyrical terms by many of the self-same writers as places of communal solidarity and neighbourliness, with all their comforts of kinship, the local and the familiar (Duffy 1985). Kavanagh's poem, 'From Tarry Flynn', is a celebration of going to help at a threshing: 'On an apple-ripe September morning/Through the mist-chill fields I went/With a pitch-fork on my shoulder/Less for use than for devilment' (1964: 28). Similarly, Sam Hanna Bell depicts the vibrant rural landscapes of Northern Ireland:

> In the distance was heard the chanting of pipes and a harmonious murmur of voices. In the owl-light there appeared over a rise in the road the piper followed by twenty or thirty lads and girls. Some of them, arm-in-arm, were prancing before him as he played, others, weary-footed, trailed behind him, and the rest, on wavering slowly moving bicycles, brought up the rear.
>
> (Bell 1974: 207)

Sean Ó Faoláin, albeit a fierce critic of the narrow vision of Ireland's rural ideology, displays something of the same ambivalence as Kavanagh in the strong connections which he had with his home landscape by the Deel in Limerick; 'that sighing land, wet above and wet underfoot and that season between the last threshing and the first ploughing that became his land and his season, a world of brown hayricks wind tumbled' (quoted in O'Connor 1996: 21).

Despite the propensity of Irish writers to evoke the mysticism of arcadia, many readings of rural Ireland unambiguously depict the social claustro-phobia and oppressiveness in these countrysides. MacNamara's *Valley of the Squinting Windows* (1918) was highly controversial because of its clear assault on the rural ethos. To Frank O'Connor (1950: 269), the sentiments expressed in William Allingham's mawkish verse on Ballyshannon – 'the kindly spot, the friendly town, where everyone is known' – 'send shudders down my spine'. 'I must get out of here . . . out of Ireland . . . I've had enough of it, it's all down on top of you. Like a load of hay. There's no space here. No scope. It's too small', said one of Ó Faoláin's characters (1940: 122). Sam Hanna Bell's *December Bride* (1974) points to a similar oppressiveness in Presbyterian County Down. Such negative renditions of the rural myth survived into the 1960s, despite coinciding with the Republic's first economic and industrial revolution. For example, John McGahern's *The Dark* (1965) – banned on publication – was written against the bleak, rush-infested landscape of Leitrim and the north-west, Ireland's 'black-hole' of emigration for more than a century, and an area which epitomised the social despair of the rural world. Both McGahern and Edna O'Brien (1960) commemorated the struggle of individuals – young people, women, lonely souls – to escape the suffocating grip of fields and family in the countryside.

This preoccupation with both positive and negative renditions of rurality reflects the simple point that, until the 1960s, much of Ireland – North and South – was a predominantly rural society. Many of its most influential writers were from the countryside and, inevitably, rural imagery coloured their writing. Even today, Seamus Heaney constantly reaches for rural metaphors while speaking powerfully of the importance of 'ancestral worlds' – invariably rural – as a means of awakening us to the reality of our world and its past (Heaney 1990, 1993). Again, Brian Friel's play, *Ballybeg* (1984), is a territorial metaphor through which he examines three decades of change in rural Ireland. Is it surprising that in a letter to *The Irish Times* (14 August 1996) a visitor vented what may well be a more general frustration?: 'In this large capital of a significantly urbanised country, why do the best play-wrights explore national identity in exclusively rural terms?' For some people in Ireland today, rural life and landscape is a fading memory; for most it may be more myth than reality.

THE BIG HOUSE

The countryside, moreover, was also the setting for other renditions of Irish identity. Themes and images revolving round the 'big house', landlords and landed estates form a not unexpected element in Irish literature from the nineteenth century onwards. The socio-economic and political problems of the country were increasingly linked to the tensions aroused by the land issue, the social pre-eminence of a powerful landowning class being seen by some as the cause of the problems, by others as the means to a solution. Maria Edgeworth's *Castle Rackrent* presented a model of Irish problems with solutions which would introduce civility into Ireland's affairs. Within the parameters of the expectations of a largely English readership, she portrayed the wild, undisciplined and drunken behaviour of the Rackrents and their followers. Edgeworth's 'dream of order relied for its foundation upon an alliance of the wild Irish aristocratic type (Catholic or Protestant) with the pragmatic English spirit' (Deane 1987: 104). The only future she saw lay in the introduction of British manufactures, or the return of all absentee landlords. In Carleton's *Traits and Stories*, while the peasant landscapes remain dominant, the big house exists between the lines and his later novels were written as prescriptions for change in which the estate played a key redemptive role. These views fitted with prevailing English attitudes. William Steuart Trench's report on the problems of the Shirley estate in County Monaghan (1843) emphasises the importance for the landlord of encouraging and helping tenants of 'good character' to improve their conditions: 'The slated houses dotted over the property, the individual substantial tenants who have each of them felt some proof of the landlord's kindness . . . stand as so many rocks to stem the popular tumult.' These essentially élitist and conservative representations of Ireland, in which the landowning class was perceived as holding the key to future social stability, gradually changed after the mid-century. This reflected the economic decline of the estates and the political and social impoverishment of the élite through measures like the Disestablishment of the Church of Ireland, the Land Acts and 1898 Local Government Act. The Land Acts especially, which dismantled the estates, signified the beginning of the end of the big house and the life it contained.

Representations of the big house in Irish landscape and society occasionally capture the schizophrenic nature of identity among many of the Anglo-Irish landed class, a characteristic feature which emerged more starkly as the century progressed and cultural and political polarities became more obvious. The social world of the Irish landed élite was divided between England and Ireland – many intermarried with English landed families and many were educated in England. There was a tradition from the seventeenth century that at the first sign of trouble in Ireland, they headed back to England, like the Countess of Mayo who found a man hiding under her bed in 1798 and sailed to England never to return. Following the union, there was a significant seasonal movement between houses in Ireland, where ennui was an endemic affliction for the residents, and England (Somerville-Large 1995). Among twentieth-century writers, Elizabeth Bowen – living on the hyphen between 'Anglo' and 'Irish' (Kiberd 1995: 368) – was a fitting retrospective witness to this confusion of identity, most notably in *The Last September* (1948). For her class, only really at home between Holyhead and Dún Laoghaire, the Irish landed estate was always something between a '*raison d'être* and a predicament' (quoted in Kiberd 1995: 376) where, as Lady Fingall noted, the Anglo-Irish lived in 'a world of their own with Ireland outside the gates' (Somerville-Large 1995: 355).

On the other hand, as an antidote to these particular quandaries of identity, many among the landowning class evoked a strong sense of Irish place, its naming and claiming being one of the distinguishing features of the Anglo-Irish writing tradition. This is epitomised in Lady Gregory's evocation of Coole on the eve of its loss in 1920:

> Coole has been a place of peace . . . a home of culture in more senses than one. Arthur Young found Mr Gregory making a 'noble nursery the plantations for which would change the face of the district' and those woods still remain; my husband added the rare trees to them and I have added acres and acres of young wood. Richard Gregory collected the fine library . . . Robert loved [Coole] and showed its wild beauty in his paintings . . . and through the guests who have stayed there it counts for much in the awakening of the spiritual and intellectual side of our country.
>
> (Gregory 1946: 15)

However, Kavanagh's condemnation of the destruction of the trees on the Rocksavage estate at much the same time is an apposite comment on what was often the hostile new world of the big house – 'There was no love for beauty. We were barbarians just emerged from the Penal days' (1971: 63; Duffy 1985). The house itself, probably the most significant landscaping achievement of the Anglo-Irish, features prominently in many of Molly Keane's novels as a setting for the lives of its occupants that is often disconnected from the other Irelands beyond its gates. In *Full House*:

> It was dark and flat and just too high in the middle like some early Georgian houses are, but its wings were of such extreme grace and proportion that their steepness was a welcome and faintly acid contrast to their inevitable correctness . . . . At Silverue there were two round halls opening one out of the other and a double, twining staircase of lovely swinging curves – two airy curves perfectly realised in wood. A romantic staircase, perpetually pleasing.
>
> (Keane 1986: 21)

Yeats's love affair with the house at Coole, celebrated in his poem 'Coole and Ballylee' (1931), was probably outdated and misdirected, a product of his association with the endeavours of Lady Gregory, his own Anglo-Irish background and a belief in the possibilities of bridging the continuing chasm between the estate and the peasant. In 'A Vision', however, he castigated the philistinism of the new state which rejected the legacy of the big house: 'To kill a house/ Where great men grew up, married, died,/ I here declare a capital offence' (Yeats 1989: 683).

Thompson's *Woodbrook* (1974) and Farrell's *Troubles* (1970) capture well the crumbling fortunes of the increasingly marginalised world of the big house in the early twentieth century, as genteel owners tried to hang on to the certainties of empire which were slipping away in the chaos of world war and its aftermath and in the social democracy of a war of independence. While Yeats foresaw the destruction of Coole, 'when nettles wave upon a shapeless mound', Molly Keane employs the image of a leaking roof to symbolise the more characteristic fate of the big house and its twilight atmosphere of decline: 'the sound of a drip of water, dripping from the roof, falling all the way down the beautiful hollow height of the house, falling like a body through the air and landing punctually as the tick of a clock in a tin basin at the stair foot' (cited in Somerville-Large 1995: 356).

Although late twentieth-century Ireland has claimed as its own its distinguished architectural and cultural inheritance of big houses – a result of both the heritage industry and historical revisionism – some of the ambivalences in representations of the big house and its symbolism for Ireland remain. These are well captured in Michael Hartnett's poem, 'A Visit to Castletown House' (a useful symbol as the headquarters of the Irish Georgian Society and now the property of the state):

> The house was lifted by two pillared wings
> out of its bulk of solid chisellings
> and flashed across the chestnut-marshalled lawn.
> [He goes outside]
> into the gentler evening air
> and saw black figures dancing on the lawn,
> Eviction, Droit de Seigneur, Broken Bones:
> and heard the crack of ligaments being torn
> and smelled the clinging blood upon the stones.
>
> (Hartnett 1994: 45–6)

THE URBAN WORLD

There is little that is new in the opposition of city and country as social and cultural constructs in Irish literary representations. The urban was elided in many of the idealisations of the rural from Yeats onwards – although Kavanagh's 'malignant Dublin' may have been a more ambiguously personal reflection. Indeed, rejection of the city, its secularism and its bourgeois compromises on cultural nationalism, is implicit in the construction of the rural idyll in Ireland. Historically the headquarters of British authority in Ireland, Dublin was suspect on the national issue in the early twentieth century. It was middle-class Dublin Catholic society which so strongly opposed Yeats and the Abbey Theatre, and whose attitudes contributed to the exile of writers like Joyce and, later, Séan O'Casey. As the only city, apart from Belfast, with a substantial working class, which became involved in the labour movement of the early twentieth century before ultimately forming an uneasy alliance with cultural nationalism, it had little enough in common with idyllic rural Irelands of whatever hue (Gibbons 1996: 95). Not surprisingly, the city has reciprocated rural animosities. Donagh MacDonagh's poem, 'Dublin made me', for example, mockingly claims: 'Dublin made me and no little town/With the country closing in on its streets/The cattle walking proudly on its pavements/The jobbers, the gombeenmen and the cheats/Devouring the fair-day between them' (cited in Longley 1992: 36).

As a middle-class Dubliner, James Joyce had a limited understanding of the countryside. *Dubliners* is a modernist representation of an urban society in all its sordidness and mediocrity, far removed from the heroic rural world of Yeats and Synge and the Celtic Revival. However, one story – 'The Dead' – does focus on the antagonism between city and country in Ireland, especially as expressed in the then emergent cultural nationalism. In it, Miss Ivors tries to persuade Gabriel Conroy to join a group excursion to the Aran Islands. He claims to be more interested in visiting France or Belgium, partly to keep in touch with the languages and partly for a change:

'And haven't you your own language to keep in touch with – Irish?'
'If it comes to that, you know, Irish is not my language'. . . .
'And haven't you your own land to visit' continued Miss Ivors, 'that you know nothing of, your own people, and your own country?'
'O, to tell you the truth' retorted Gabriel suddenly, 'I'm sick of my own country, sick of it'. . . .
She looked at him from under her brow for a moment quizzically until he smiled. Then . . . she stood on tiptoe and whispered into his ear: 'West Briton!'

(Joyce 1993b: 102)

At the end of the story Conroy tries to acknowledge the connection between Dublin and the West with the famous unifying image of snow falling ' . . . all over Ireland. It was falling on every part of the dark central plain, on the treeless hills, falling softly on the Bog of Allen and, farther westward, softly falling into the dark mutinous Shannon waves' (Joyce 1993b: 120).

Joyce's portraiture of Dublin and its people was his greatest artistic achievement. His writing continues to be an exemplary representation of the landscape, moods and busyness of a city. The seediness of early twentieth-century Dublin's red-light district, located among some of the most notorious slums in Europe, is authentically represented in *A Portrait of the Artist as a Young Man*:

He had wandered into a maze of narrow and dirty streets. From the foul laneways he heard bursts of hoarse riot and wrangling and the drawling of drunken singers. He wandered onward, undismayed,

75

wondering whether he had strayed into the quarter of the Jews. Women and girls dressed in long vivid gowns traversed the street from house to house.

(Joyce 1993a: 174)

Joyce also captures the ambience of the cityscape and the sense of place of the built environment through which his characters constantly move. *Ulysses* is essentially an urban odyssey:

By lorries along Sir John Rogerson's Quay Mr Bloom walked soberly, past Windmill Lane, Leask's the linseed crusher's, the postal telegraph office . . . the sailors' home. He turned from the morning noises of the quayside and walked through Lime Street. By Brady's cottages, a boy for the skin lolled, his bucket of offal linked, smoking a chewed fagbutt.

(Joyce 1993c: 305)

Again, Séan O'Casey's descriptions of the city, as witnessed from inside its tenement slums, were flavoured with a strong social awareness of injustice, the ubiquitous presence of disease and death and a marked disenchantment with the quintessentially rural ideology of the Irish Free State:

He could see the street stretching along outside, its roughly cobbled roadway beset with empty match-boxes, tattered straws, tattered papers, scattered mounds of horse-dung, and sprinkled deep with slumbering dust . . . . Lean-looking gas-lamps stood at regular intervals on the footpaths . . . by the side of the tall houses, leading everywhere to tarnishing labour, to consumption's cough, to the writhings of fever.

(O'Casey 1949: 57)

Brian O'Nolan (Flann O'Brien) described Dublin during the depressed 1940s and 1950s, an inward-looking city, closed in by war and independence. He was foremost among a coterie of artistic spirits striving against the narrowness of Irish life, the authoritarianism of bishops and the ideological aridity of government – 'the cheering nation stagnant' in Dermot Bolger's retrospective impression of 1958 in the city (1986: 41). Dublin, its slowly expanding civil service supplied from the countryside, remained eclipsed by the 'sanctification of rural Ireland' promulgated by this self-same officialdom (Kiberd 1995: 495). As Flann O'Brien (1967: 89), O'Nolan was at his most caustic, however, on the rural myth – 'a wondrous glen it is with green-streamed water, containing multitudes of righteous people and a synod of saints' – and set out on his mission to overturn all the mythic clichés of Gaelic Irishdom.

By the 1980s, however, writers like Roddy Doyle (1992) and Dermot Bolger (1990) had begun – very ambiguously – to address Dublin's maturity as a western capital city. This seems no less than appropriate as the city accounts for nearly one-third of the population of a state ready to break out of its imprisonment in ideologies of identity vested in rurality. 'Dublin 4' is geographical shorthand for a mind-set that epitomises the image of middle-class, liberal, urban society. But the city of Bolger and Doyle reflects the ills of western capitalism rather more than a 'vibrant zone of creativity' (Kiberd 1995: 95). It is the converse of the dark secrecies of the rural world. There is no place to hide in the city where chronic unemployment, drug addiction, domestic violence and environmental degradation flourish in the working-class suburbs, where 'ambulances spurt blue flames down shrunken passageways' and 'stolen cars zigzagged through the distant grey estates' (Bolger 1986: 27; 1990: 35). Significantly, surveys in the rural west of Ireland during the 1980s also reflected this sense of alienation. Unlike Edna O'Brien's magnetic city of the 1950s and 1960s, Dublin is now a place to be avoided by those who live in an infinitely preferable rural world (Breathnach 1984: 66). The Dublin–Ireland dichotomy remains sharply defined by social issues like divorce and abortion.

THE NORTH

In the context of emerging national consciousnesses in Ireland, the North has inherited intractable problems of identity which are well reflected in its literature. Cultural nationalism in the South and its emphasis on Gaelic and Catholic values excluded a significant section of Ulster's population from dominant images of Irishness. Literature in the North reflects the ensuing confusions of identity and cultural uncertainties, best exemplified by poets like John Hewitt and Derek Mahon, whose work has been described as portraying the 'spiritual desolation' of Ulster Protestant culture (Longley 1984: 17). In his post-Partition search for identity, Hewitt expressed the problem thus: 'People of Planter stock often suffer from some crisis of identity, of not knowing where they belong . . . some call themselves British, some Irish, some Ulstermen, usually with a degree of hesitation or mental fumbling' (Hewitt 1987: 122). This is not unlike Seamus Heaney's 'doubleness of focus' in Irish, but especially Northern Irish, identity. Graham (1994: 266) suggests that Protestant identity in Northern Ireland has 'no text of place as provided by a representative landscape', the result being a negative, defensive identity, defined in opposition to the Otherness of the threatening Gaelic and Catholic South. Protestant identity is principally a politically grounded ('British') phenomenon, with a historical iconography emphasising a siege mentality, which excludes the Catholic minority in the North. Heaney describes his early awareness of this 'disjointed Ulster':

It was a large map of this younger, smaller Ulster that hung in different shades of greens and blues and fawns in the first classroom I knew, with the border emphasised by a thick red selvedge all the way from Lough Foyle to Carlingford Lough. That vestigially bloody marking halted the eye travelling south or west; but travelling east, on slender dotted lines that curled fluently from Larne to Stranraer, from Derry to Glasgow . . . small black steamships lured the eye across the blue wash of the North Channel. Another emblem there. . . .

(Heaney 1983: 18)

The literature of Northern Ireland inevitably reflects these nuances of contested identity. For Hewitt, it meant seeking a shared heritage in Ulster, which never quite developed beyond being a regional subset of an Irish cultural identity. Like other poets in Northern Ireland, he became interested in the possibilities of territory and landscape as a nexus for cultural identity, on the basis that the landscape is a shared legacy with which all sections in the community can identify. In somewhat Yeatsian terms, Hewitt suggested that the Ulster writer 'must be a *rooted* man, must carry the native tang of his idiom like the native dust on his sleeve; otherwise he is an airy internationalist . . . ' (1987: 115). In the end, however, Hewitt's writing illustrates his continuing search for identity and the use of words in his writing like 'roots', 'conacre', 'freehold', 'place', 'colony', 'planter stock' might be taken as reflective of a settler society in search of its place. In the poem 'Conacre', published in 1943, he expresses the limitations of this search, commitment to his incomplete region being expressed in the recurring spatial imagery of occupying a ledge: 'This is my home and country. Later on /perhaps I'll find this nation is my own/ but here and now it is enough to love/ this faulted ledge' (1991: 9–10).

Apart from the more obvious cultural contrasts between Northern Ireland and the Republic, one abiding difference, which has undoubtedly served to separate the historical experience of the two parts of the island, is the greater degree of urbanisation in the North. The urban–industrial experience of east Ulster especially was akin to that of northern England or central Scotland. It is probably significant, therefore, that in reflecting this urban perspective, many Ulster writers have emphasised their problems of identifying with an Ireland that in real and representative terms was rural. Hewitt, the middle-class urbanite, showed some lack of understanding of rural Ireland in one minor poem: 'The hospitable Irish/ come out to see who passes/ bid you sit by the fire/ till it is time for mass' (1991: 150). Derek Mahon is also more at home articulating the world of the city and town, suburban Belfast especially – 'I lived there as a boy and know the coal/ Glittering in its shed, late-afternoon/ Lambency informing the deal table,/ The ceiling cradled in a radiant spoon' (Mahon 1991: 120). Playwrights such as Graham Reid and Stewart Parker have depicted the Protestant working-class landscapes of Belfast, representations of life which – in the midst of all the conflict – still present a fairly uncomplicated world-view, a self-contained people, confident in their territory and numbers, certain about their belonging and their community. But a poem by Hewitt in 1935 reflected on the latent violence in ethnic urban territoriality, marked to the present in symbols like marches, graffiti and flags: 'in the city of our dreadful night/ men fought with men because of a threadbare flag/ or history distorted in temper' (Hewitt 1991: 16).

Graham (1994: 275) summarises the multifaceted nature of Northern Irish identity in which the unifying potential of invented landscapes is lost because ultimately they are contested landscapes. Nowhere is this more evident than in the borderlands where both communities interface, whether in the micro-geography of the cities, or in the equally finely divided countryside. The tensions in the south Ulster borders, where the landscape is intimately divided into Protestant and Catholic farms and townlands, have been well expressed by the Monaghan writer Eugene McCabe:

> For Canon Leo McManus the best part of his ministry was travelling on horseback the by-roads, farms, villages and townlands of Upper Fermanagh. On the walls of his dining-room he had land commission maps pencil-marked, and could tell at a glance the name, status and religion of the owner . . . . The county was evenly divided between Catholic and Protestant; time and determination, God willing, would alter that.
>
> (McCabe 1992: 19)

ON EMIGRATION AND EXILE

While the population of Ireland has fallen constantly since the 1850s, its towns – Belfast and Dublin excepted – have played only a minor role in intercepting the demographic exodus from the land. Part of the myth of nationalism in Ireland has been a form of denial of the reality of emigration, especially the trek from rural arcadia to the urban metropolises of Britain and the New World. De Valera's vision of Ireland, for instance, was spelled out coincidentally with some of the heaviest out-migrations; both church and state extolled the virtues of rural, pastoral Ireland and condemned the social instability (the 'sinfulness') of the city, especially the English city. Simultaneously, the future survival of the rural communities seemed to depend on the absorption of their surplus members by these cities, 'shunted in a swaying tube/ to a dour migrant workers' hostel' (Bolger 1986: 74).

As suggested already, many of the literary representations of rurality and of the West in myths of nationalism were urban in origin. It was, however, precisely the rural areas of the West – the mythic land of Yeats, Synge and Pearse – which provided most emigrants. Paul Durcan (1975: 63) represents the continuing bitterness of many who were forced to emigrate in the 1880s in spite of the pieties of the old Ireland: 'She was America-bound, at summer's end/ She had no choice but to leave her home –/The girl with the keys to Pearse's cottage.' MacGabhan (1973: 138) recalls that in the 1890s, 'you'd think there must be no one left in Cloghaneely [County Donegal] – that they were all in New York.' Many Irish writers have expressed the disillusionment with rural life which drove so many to emigrate. Indeed, Joyce, O'Casey, Beckett, Frank O'Connor, Kavanagh, Edna O'Brien and many others were either temporary or permanent emigrants themselves, their personal experiences authenticating the nature of emigration as a reality for this society. O'Casey was but one who fled from the suffocating dominance of the 'priest's cassock and the friar's gown' in the life of the state in the 1930s. Undoubtedly, much literature reflects ordinary migrants' dissatisfaction with the repressive role of Catholicism in the Irish Free State, where dancing, cinema and courtship were strictly chaperoned by the mullahs of the triumphalist church.

O'Brien and Ó Faoláin, however, also capture the simpler desire of many to escape to the bigger urban world where economic and social opportunities were greater. Edna O'Brien (1960: 168) refers to the country girl's delight in the 'neon fairyland of Dublin . . . . I loved it more than I had ever loved a summer's day in a hayfield. Lights, faces, traffic, the enormous vitality of people hurrying to somewhere'. Ó Faoláin's (1940: 122) character too 'almost hugged the sensation of release' on arriving in New York. Richard Power (1980: 197) refers to the vitality of Birmingham in the 1950s: 'in the pubs, lights were gleaming, the clash of glasses, high-pitched conversations, the warmth of fellowship'. In the 1990s, similar sentiments are still being expressed by writers like Dermot Bolger, Joseph O'Connor and Emma Donoghue.

The other great paradox in Irish emigration is the motif of exile. Miller (1985) suggests that emigration-as-exile was a social strategy to salve the pain of emigration for both emigrants and the rural communities unable to hold on to them. Much of the rich legacy of songs and ballads on emigration evokes the theme of reluctant exile from Ireland to roam in foreign lands. According to Deane (1994: 34), 'exile, the high cultural form of emigration, became one of the most favoured strategies for the representation of Ireland in the early twentieth century. It was a form of dispossession that retained – imaginatively – the claim to possession.' Irish emigrants became, as it were, more Irish than the Irish themselves. In many ways, the ultimate idea of Ireland and what it represented to those who remained at home was invented by those who had to leave (Kiberd 1995: 2).

CONCLUSION

Thus literature and art reinforce the notion of complex and contested representations of Irishness. Texts can be read in different ways as in the rural idyll, the embodiment of the nation-state but at the same time resonant with oppression and claustrophobia. In concluding, it might be suggested that the tourism industry, perhaps the most rapidly expanding sector in Ireland's contemporary economy, is one of the most influential forces now shaping representations of identity, landscape and culture. Largely based on an urban consumption of landscape and place, stereotypical constructions of identity form the very stuff of tourism and these are well reflected in artistic representations from paintings, photographic images and writing. Most of the themes in this chapter – the myths of the West, the opposition of rural and urban cultures, the Anglo-Irish legacy, even the northern landscapes of conflict, as well as the writers, their personalities and the landscapes they wrote about – have been appropriated by the tourism industry.

Narrative places, 'storied landscapes', literary places and landmarks are ideally suited to tourism promotion (Herbert 1995: 12). Yeats's work, for example, is so pervasive that his poetic representations of Sligo are more 'real' than the actuality. This is a common characteristic of 'writers' countries', which have been developed all over the western world by tourism: for instance, there are now at least three levels of reality in Herriot's Yorkshire – the book, the television series and the film. 'Bloomsday' in Dublin is probably the most extreme example of this appropriation of a landscape and an event that never happened. 'Yeats Country', 'Kavanagh Country' and 'Goldsmith Country' are some of the new territorial designations which have been promoted by tourism and local authorities in the hope of luring visitors to experience the landscape and places which inspired or were celebrated by writers. In some cases, the writers provide the focus for specific developments, such as the Patrick Kavanagh Literary and Rural Resource Centre in Iniskeen, County Monaghan, or the Clogher Valley Rural Centre, based in William Carleton's country in Tyrone. An expanding array of summer schools take advantage of the opportunities to sell Yeats, Hewitt, Carleton, Joyce, Kate O'Brien, Patrick McGill, Jonathan Swift, Oliver Goldsmith, Brian Merriman and many others as cultural tourism products. It is through literature and its readings, as well as its geography, that Irish place is defined and redefined, constantly negotiated as society is contested along its many and varied axes of differentiation by its myriad actors and their conflicting motivations.

REFERENCES

Barnes, T. J. and Duncan, J. S. (eds) (1992) *Writing Worlds: Discourse, Text and Metaphor in the Representation of Landscape*, London: Routledge.
Bell, Sam Hanna (1974) *December Bride*, Belfast: Blackstaff.
Bolger, D. (1986) *Internal Exiles: Poems*, Mountrath: Dolmen Press.
—— (1990) *The Journey Home*, London: Viking.
Bowen, E. (1948) *The Last September*, London: Cape.

Breathnach, P. (1984) *Community Perceptions of Development and Change in the Conamara Theas and Corca Dhuibhne Gaeltacht*, Maynooth: Department of Geography.

Brown, M. (1972) *The Politics of Irish Literature*, Seattle: University of Washington Press.

Brown, T. (1981) *Ireland: A Social and Cultural History 1922–1985*, London: Fontana.

Carbery, M. (1938) *The Farm by Lough Gur*, London: Catholic Book Club.

Carleton, W. (1865) *Traits and Stories of the Irish Peasantry*, London: William Tegg.

Cloke, P., Philo, C. and Sadler, D. (1991) *Approaching Human Geography*, London: Paul Chapman.

Cosgrove, M. (1995) 'Paul Henry and Achill Island', in U. Kockel (ed.) *Landscape, Heritage and Identity: Case Studies in Irish Ethnography*, Liverpool: Liverpool University Press.

Deane, S. (1987) 'Irish national character, 1790–1900', in T. Dunne (ed.) *The Writer as Witness*, Cork: Cork University Press.

—— (1994) 'Land and soil – a territorial imperative', *History Ireland* 2, 1: 31–4.

Doyle, R. (1992) *The Barrytown Trilogy*, London: Secker and Warburg.

Duffy, P. J. (1985) 'Carleton, Kavanagh and the south Ulster landscape, c. 1800–1950', *Irish Geography* 18: 25–37.

—— (1994) The changing rural landscape, 1750–1850: pictorial evidence', in B. Kearney and R. Gillespie (eds) *Ireland: Art Into History*, Dublin: Town House.

—— (1995) 'Literary reflections on Irish migration in the nineteenth and twentieth centuries', in R. King, J. Connell and P. White (eds) *Writing Across Worlds: Literature and Migration*, London: Routledge.

Dunne, T. (ed.) (1987) *The Writer as Witness*, Cork: Cork University Press.

Durcan, P. (1975) *O Westport in the Light of Asia Minor*, Dublin: Anna Livia Books.

Edgeworth, M. (1964 World Classics ed.), *Castle Rackrent*, Oxford: Oxford University Press.

Farrell, J.G. (1970) *Troubles*, London: Fontana.

Foster, R. (1989) 'Varieties of Irishness', in M. Crozier (ed.) *Cultural Traditions in Northern Ireland*, Belfast: Institute of Irish Studies.

Friel, Brian (1984) *Selected Plays,* London: Faber.

Gibbons, L. (1996) *Transformations in Irish Culture*, Cork: Cork University Press/Field Day.

Graham, B. J. (1994) 'No place of the mind: contested Protestant representations of Ulster', *Ecumene* 1, 3: 257–81.

Gregory, Lady Augusta (1946) *Lady Gregory's Journals, 1916–1930*, London: Putnam and Co.

Hartnett, M. (1994) *Selected and New Poems*, Oldcastle: Gallery Press.

Heaney, S. (1980) *Preoccupations: Selected Prose 1968–78*, London: Faber.

—— (1983) 'Forked tongues, *céilis* and incubators', *Fortnight* 18 September.

—— (1990) 'Correspondences: emigrants and inner exiles', in R. Kearney (ed.) *Migrations: The Irish at Home and Abroad*, Dublin: Wolfhound Press.

—— (1993) 'The sense of the past', *History Ireland* 1, 4: 33–7.

Herbert, D. T. (ed.) (1995) *Heritage, Tourism and Society*, London: Mansell.

Hewitt, J. (1991) *The Collected Poems of John Hewitt*, ed. F. Ormsby, Belfast: Blackstaff Press.

Joyce, J. (1993a) *A Portrait of the Artist as a Young Man*, London: Chancellor.

—— (1993b) *Dubliners*, London: Chancellor.

—— (1993c) *Ulysses*, London: Chancellor.

Kavanagh, P. (1948) *Tarry Flynn*, London: The Pilot Press.
—— (1964) *Collected Poems*, London: MacGibbon and Kee.
—— (1971) *The Green Fool*, London: HarperCollins Publishers.
Keane, J. B. (1966) *The Field*, Dublin: Mercier.
Keane, Molly (M.J. Farrell) (1986) *Full House*, London: Virago.
Kiberd, D. (1995) *Inventing Ireland: The Literature of the Modern Nation*, London: Cape.
Leerssen, J. (1994) 'The western mirage: on the Celtic chronotype in the European imagination', in T. Collins (ed.) *Decoding the Landscape*, Galway: Centre for Landscape Studies.
Livingstone, D. (1992) *The Geographical Tradition*, Oxford: Blackwell.
Longley, E. (1984) 'An ironic conscience at one minute to midnight', *Fortnight* 17 December.
—— (ed.) (1992) *Culture in Ireland: Division or Diversity?*, Belfast: Institute of Irish Studies.
MacGabhan, M. (1973) *Rotha Mór an tSaoil*, (translated as *The Hard Road to Klondyke*), London: Routledge & Kegan Paul.
MacNamara, B. (1918) *The Valley of the Squinting Windows*, London: Maunsel and Co.
Mahon, D. (1991) *Selected Poems*, Harmondsworth: Viking Press in association with Oxford University Press.
Mason, E. and Ellmann, R. (1959) *The Critical Writings of James Joyce*, New York: Viking Press.
McCabe, E (1992) *Death and Nightingales*, London: Secker and Warburg.
McGahern, J. (1965) *The Dark*, London: Faber.
Miller, K. A. (1985) *Emigrants and Exiles: Ireland and the Irish Exodus to North America*, Oxford: Oxford University Press.
Ó Faoláin, S. (1940) *Come Back to Erin*, London: Cape.
O'Brien, E. (1960) *The Country Girls*, London: Hutchinson.
O'Brien, F. (1967) *At-Swim-Two-Birds*, London: Penguin.
O'Casey, S. (1949) *Inishfallen Fare Thee Well*, New York: Macmillan.
O'Connor, F. (1950) *Leinster, Munster, Connaught*, London: Robert Hale.
O'Connor, P. (1996) *All Ireland Is In and About Rathkeale*, Newcastle West: Oireacht na Mumhan Books.
Pearse, P.H. (1924) *The Collected Works of Padraic H. Pearse*, Dublin: Phoenix Press.
Power, R. (1980) *Apple on a Tree-top*, Dublin: Poolbeg.
Relph, E. (1976) *Place and Placelessness*, London: Pion.
Sheerin, P. (1994) 'The narrative creation of place: the example of Yeats', in T. Collins (ed.) *Decoding the Landscape*, Galway: Centre for Landscape Studies.
Somerville-Large, P. (1995) *The Irish Country House: A Social History*, London: Sinclair-Stevenson.
Synge, J. M. (1910) *The Work of John M. Synge*, Dublin: Maunsel and Co.
Taylor, A. (1988) *To School Through the Fields*, Dingle: Brandon Books.
Thompson, D. (1974) *Woodbrook*, London: Beaver and Jenkins.
Trench, W. S. (1843) *Report on the Shirley Estate*, Public Record Office of Northern Ireland, D3531/S/55.
—— (1868) *Realities of Irish Life*, London: Longman.
Whelan, K. (1993) 'The bases of regionalism', in P. Ó Drisceoil (ed.) *Culture in Ireland: Regions, Identity and Power*, Belfast: Institute of Irish Studies.
Yeats, W. B. (1982) *Collected Plays*, Dublin: Gill and Macmillan.
—— (1989) *Yeats's Poems*, ed. A. N. Jeffares, Dublin: Gill and Macmillan.

Part II

AXES OF DIVISION AND INTEGRATION

INTRODUCTION

The starkly opposed linear narratives of Irish identity were a product of the internal dynamics of social and political change in the island during the late nineteenth and early twentieth centuries. But they have also been reinforced by tourism imagery and by external perceptions of Irish life in which it is all too easy to identify what one commentator has referred to as 'intellectual disengagement'. The continuing acceptance beyond Ireland of stereotypical images of Irishness and its place – inevitably sympathetic to the nationalist ethos while condemnatory and patronising of the dourness of Ulster Protestant unionism – points to a widespread suspension of intellectual judgement on Ireland and tacit support of political, social and cultural forms that would not be countenanced so uncritically elsewhere in Europe.

The purpose of Part II is to redress such narrowness of vision by examining something of the complexity of identities and attitudes that have been subsumed by the reconstruction of Irishness around the national issue. In Chapter 5, Peter Shirlow describes how class interests in both North and South were subordinated within 'devoted national economies' which proclaimed state-directed fealty to all classes, economic allegiance being constructed as a parallel to nationalism. In reality, the winners were the indigenous, intensely conservative and Catholic capital-owning middle classes in the South and the unionist élite in the North operating through structures of paternalised sectarianism. The chapter explores the contemporary repercussions of these past alignments, emphasising the widespread differentials of income and opportunity which characterise the contemporary South and the intra-working-class war that is the northern conflict. Meanwhile, British policy in the North has ensured the expansion of an accommodationist middle class – both Catholic and Protestant – which ultimately is more likely to embrace limited all-Ireland institutions.

The 'cause' for or against British involvement in Ireland has also been successful in marginalising issues of gender and sexual inequality. In Chapter 6, Catherine Nash demonstrates how the gendering of Ireland

and the construction of nationalist versions of ideal Irish femininity have been deeply problematical and damaging for women. The subordination of women, depicted as passive and domestic, desexualised and venerated as 'mother', was ratified in the 1937 Constitution, which also perpetuated the notion of heroic masculine sacrifice for the idealised, allegorical mother-figure of Ireland. The chapter goes on to demonstrate how women in Ireland are now attempting to reconceptualise ideas of nation, national traditions and relationships to place in more open and inclusive ways. Gender is being renegotiated through political action in Northern Ireland, while the heterosexism of traditional nationalism is also being challenged.

Ethnicity provides a final axis of differentiation, albeit one which moves the discussion closer to more conventional nationalistic terms. In Chapter 7, Mike Poole argues that the ways in which we conceptualise Irish society are profoundly important for the ways in which we think about its mutually antagonistic groups. He argues that socially exclusive ethnicity provides the essence of the northern conflict but also that ethnic attachment is a variable feeling, subject to ebb and flow. He sees little evidence for a de-ethnicised bloc in Northern Ireland, but instead a whole series of issues on which people may be de-ethnicised. In the Republic, conversely, there is considerable evidence of a decline in ethnicity as nationalism transforms along an axis from ethnic criteria of exclusion to civic criteria of inclusion.

5

CLASS, MATERIALISM
AND THE FRACTURING
OF TRADITIONAL ALIGNMENTS

Peter Shirlow

INTRODUCTION

Ireland's traditional socio-economic and class profiles have altered dramatically since partition. In both the Republic and Northern Ireland, socio-cultural indicators such as the emergence of high-technology industries, growth in disposable income and the professionalisation of certain manual trades seem to suggest that both societies have merged fortuitously into the contemporary globalising economy (Foley and McAleese 1991). Furthermore, the two main cities, Dublin and Belfast, have been transformed into arenas of conspicuous consumption. The former is now one of the fasting growing tourism destinations in Europe, while the latter has the highest rates of profit for large-scale retailers in the UK. Moreover, since 1945, real increases in disposable income have been paralleled by meaningful social progression in the form of substantial improvements in education, health and housing provision.

Simultaneously, however, homelessness, unemployment, emigration and welfare dependence have increased dramatically. In Northern Ireland the previously powerful manufacturing base is in terminal decline, due to its long-term over-dependence on heavy industries and the failure to diversify into new markets and alternative forms of self-sustaining development (Teague 1993). Similarly, and despite its modernist drive to attract foreign capital, an under-performing indigenous sector and an over-dependence on mobile investors, foreign credit and funding from the European Union (EU) point to the Republic's economic fragility.

In exploring this dichotomy between wealth creation and social marginalisation, this chapter assesses the impact of socio-economic transition in post-partition Ireland upon class composition and modification. I argue that while both parts of Ireland tend towards convergence in terms of their socio-economic profiles, as they shift away from traditional economic forms and class alignments, the impact of socio-cultural and political change and transformation has been experienced in diverse ways. Traditional class alignments refer to the socio-economic constructs evident prior to the modernisation and transnationalisation of the Republic's economy and the commencement of de-industrialisation and civil unrest in Northern Ireland. The two traditional epochs of socio-economic activity – economic nationalism (1932–58) in the Republic, and hegemonic unionism in Northern Ireland (1921–72) – can be denoted as periods when the respective discourses incorporated allegoric perceptions of state-directed fealty to all social classes in what can be termed 'devoted' national economies. Continuing processes of socio-economic transformation have eroded these previously distinct ideological frameworks and also signalled important alterations in political and material representations of Ireland. In examining the transition from industrialism in Northern Ireland and agriculturalism in the Republic towards a more unified post-industrial/agricultural structure, I argue that while these shifts have produced similar socio-economic profiles, their repercussions have been experienced in particular and different ways throughout Ireland (Shirlow and McGovern 1996).

A CHANGING IRELAND: ECONOMY AND CLASS

The demise of traditional sources of employment has clearly impacted upon the nature of class relationships. In the Republic, agriculture, which employed half of those of working age in 1926, now accounts for a mere 11 per cent of total employment (Table 5. 1). Not surprisingly, the numerical decline of small farmer and farm labourer communities has been accompanied by a contraction of their political and material influences, underlined by the parallel processes of urbanisation, which have lessened the iconic dominance of rural Ireland in the construction of cultural and national identity.

Table 5. 1 Republic of Ireland: percentage sectoral shares of working-age population (1926/1956/1996)

Sector	1926	1956	1996
Agriculture	50	33	11
Manufacturing	12	24	23
Services	31	37	48
Unemployment	6	5	18

Source: Industrial Development Authority

In Northern Ireland, the employment share of the manufacturing sector, which at one time defined the region's particular economic role in Ireland, has declined by nearly half since 1956 (Table 5. 2). Conversely, the employment share of manufacturing virtually doubled in the Republic between 1926 and 1996, albeit from only 12 to 23 per cent. The decline in manual employment underscores a loss of industrial singularity and socio-cultural distinctiveness in Northern Ireland while, simultaneously, attesting to a form of economic confluence which renders the two economies, in terms of employment profile, increasingly similar. However, the nature of this convergence is somewhat illusory, given the nature of employment in Northern Ireland. Over half of the 62 per cent employed within the service sector, and three-quarters of its 22 per cent growth since 1956, can be attributed directly to conflict and civil disorder. Socio-political antagonism has artificially inflated the public sector through recruitment to the security forces and state bureaucracy.

Table 5. 2 Northern Ireland: percentage sectoral shares of working-age population (1926/1956/1996)

Sector	1926	1956	1996
Agriculture	27	13	5
Manufacturing	32	40	22
Services	34	40	62
Unemployment	6	6	11

Source: Northern Ireland Annual Abstracts of Statistics

The steady expansion of unemployment in both economies also affirms changes within the labour market. Due to competitive pressures, industry has continuously re-evaluated employment policies, directly altering previous labour market structures and employment trends through rising productivity levels, intensification of work and the deskilling (or even enskilling) of labour. Moreover, in placing a premium on price, quality and responsiveness to customer needs, increased global competition has promoted rapidly changing technologies, emphasising the need for flexible and adaptable organisations and work-forces. Such alterations in Ireland's socio-economic profile means that class is increasingly a negotiable construct, conditioning and explaining processes both of socio-economic and of cultural and political construction and modification (Peet 1991). Observable class divisions are markedly asserted and reproduced in terms of income, habitation and socially circumscribed consciousness and patterns of behaviour.

In order to determine the role played by social stratification in terms of economic power, it is essential to determine the ability of classes to accumulate wealth and income. Moreover, if the definition of class includes dimensions such as income, residence, social status, political identity and occupation, then it is clear that these terms refer to aggregates of people who share similar situations with respect to the economic system and relationships of production and consumption. Class – notoriously difficult to define – is generally employed as an umbrella term which permits the analysis of complex patterns of social divisions and integration (Peet 1991). In Ireland, the patterns of income inequality over the past thirty years indicate that the distribution of income has become more uneven due to the type and volume of employment and investment. A distinct schism now exists between a materially ascendant group (composed of the middle, higher-income, classes) and lower-income groups whose earnings and/or employment conditions have deteriorated.

This chapter utilises a developed two-class model, the key role being allocated to the bourgeoisie (or middle classes) who own and control wealth invested in production of goods and services. In terms of measurement, this class can be defined as Socio-Economic Groups (SEGs) 1–4 and 13 in the Northern Ireland Registrar General's classification (18 per cent of chief economic supporters of households) and Social Classes I–III in the Republic's Labour Force Survey (16 per cent of chief economic supporters of households in 1991). Although these higher-income groups are relatively heterogeneous in terms of attitude, rates of income and employment type, they have, since they are generally well educated and professional, benefited from the erosion of traditional class and cross-class alignments and the intensified capitalisation of Irish society – North and South – due to the infiltration of US, British, EU and Asian capital. This influx of foreign investment indicates both the fragmentation of the link between territorial cohesiveness and economic sovereignty, and also the virtual abandonment of previous modes of nationally directed economic allegiance. As such, the middle classes have in many instances discarded previous forms of economic nationalism in favour of economic transnationalism.

The remainder of the population comprises all other SEGs in the Registrar General's classification and Social Classes in Labour Force Surveys, together with the unemployed. This group, which includes manual workers, small farmers, routine non-manual workers and the inactive, has benefited less from socio-economic transformation. For example, the share of those whose livelihoods are dependent upon state-sponsored welfare in the Republic rose from 31.1 per cent to 48.1 per cent between 1971 and 1991 (Shirlow 1995). For lower-income workers and the unemployed, there has been little in the way of a coherent or collective response to the onslaught of social marginalisation, caused by the imposition of productivist and neo-conventional modes of accumulation (O'Hearn 1993). In Northern Ireland social marginalisation equates directly with advocacy for nationalism of one form or another and support for paramilitaries. It is within this constituency, with its curious affinity between Republicanism and Loyalism, that we can locate a more visible and emotional attachment to previous epochs of socio-national construction. As such, class divisions in Ireland are not simply reproduced through an uneven distribution of income, but are also duplicated by cultural and political attachment and everyday experiences.

THE REPUBLIC OF IRELAND: FROM SOCIAL IDEALS TO SOCIAL MARGINALISATION

Influenced by James Connolly's Marxist reconciliation of nationalism and socialism, the Easter Rising and Proclamation of 1916 combined demands for national self-determination and a heady mix of socialism, based upon the restoration of what was perceived as the Gaelic system of social communalism. The Proclamation's appeal to mobilised class consciousness, followed by a classless society, was soon subverted by the effects of civil war, political consolidation and the eventual emergence of Fianna Fáil as the predominant party in Irish politics (Porter and O'Hearn 1995; Gibbons 1996). Its adoption during the 1930s of indigenous capitalist-led development – as opposed to a socialist discourse – reflected a Republican ideology, which placed Catholicism, nationalist conformity and the consolidation of Irishness above the pursuit of socialist synthesis and class realignment. Articulated through a strategy of economic nationalism, the rationalisation of national homogeneity was secured by a populist absorption of class grievances. As Hazelkorn and Patterson (1995: 50) observe: 'What would prove ultimately much more significant in both its [Fianna Fáil's] politics of power and support was [an] ideology of development based on social harmony: of economic growth achieved with the minimum of social conflict.'

The pursuit of economic nationalism by Éamon de Valera's government after 1932 was enunciated through import substitution and protectionism. These policies, which survived until the late 1950s and early 1960s, were explicitly linked to a radical pretension that a merger of national identity and economic isolationism would serve both cross-class and sovereign interests. To reinforce this, Fianna Fáil promoted an allegory of devoted national interests, centred on an espousal of land redistribution, indigenous industrialisation and extended welfare provision (Breathnach 1985). The self-identified party of the 'plain people of Ireland' articulated a programme of economic recovery that was tied to an unleashing of nationalist sentiment and state intervention, the objective being to serve the cross-class interests of industrial workers, small farmers and an embryonic bourgeoisie (Smyth 1991).

Beyond the articulation of nationalism, limited welfare provision and the elevation of Irish identity, this supposedly paternalised economy failed, however, to tackle the emergence of consolidated land holdings, the capitalisation of agriculture due to the promotion of arable farming, the propensity of the bourgeoisie to invest surplus capital in foreign banks and the subsequent growth in unemployment and out-migration. If anything, Fianna Fáil was unable and unwilling wholeheartedly to adopt earlier republican principles of land and capital redistribution, as these would ultimately have threatened the material position of the middle classes and, in turn, nullified nationalist unity (Wickham 1985). Moreover, the Catholic church discouraged social reform because of fears that a strong and materially minded laity would challenge its power, a stance reinforced by periodic and volatile crusades against the theoretical egalitarianism of communism (Gibbons 1996; Lee 1986). Thus the linkages between spirituality – the Catholic church – and the devoted nationalism articulated through Fianna Fáil were not based upon the promotion of Celtic communalism and social equality. Instead, Catholic Ireland was secured for a property-owning bourgeoisie through the promotion of virtuous lifestyles and indigenous-led capitalism, uncontaminated by materialism and social progress for the masses (Keating and Desmond 1993).

The adoption of protectionism and the parochial application of economic nationalism meant that the nexus of production–consumption was founded upon a static consumer goods market, which blunted the emergence of a high-value-added market and real wage growth. The inability to develop a national version of the Fordist mass production–mass consumption complex, emerging in the 1950s throughout the western world, obstructed the creation and augmentation of material prosperity and stimulated significant out-migration. By 1958, industrial wages were 50 per cent lower than in Denmark and Britain, and 80 per cent lower than in the USA (Lee 1989). Moreover, by this time, unemployment had trebled and over 700,000 people had migrated since partition.

The failure to promote a socio-economic regime capable of satisfying the material demands of a significant section of an electorate increasingly influenced by the prosperity evident in the USA and – though to a lesser extent – Britain, as well as among those who had migrated, deepened the crisis facing the Irish state and underlined the need for alternative modes of accumulation and consumption (Smyth 1991). Economic stagnation also posed ideological constraints upon Fianna Fáil. In reaction to several electoral defeats and emerging social discord, the party was forced to make a policy choice between the continuation of economic nationalism, which was unproductive, unsustainable and unpopular, and the stimulation of investment that might promote real economic growth but in so doing would permit the importation of foreign capital (Breathnach 1995). While the latter option would ultimately dilute the economic sovereignty so central to the discourse of devoted nationalism, Fianna Fáil was clearly in danger of losing its populist base if it continued to pursue this agenda, which was engendering material frugality against the dictates of an essentially capitalist rationality of efficiency, competitiveness and extended material prosperity. The renunciation of economic nationalism, which accelerated following the replacement of de Valera by Séan Lemass in 1957, was underpinned by an alternative populist notion that economic redirection could provide social mobility and material affluence. In order to provide support for the ending of the distinct nationalist discourse, Fianna Fáil had to ensure that a process of transnationalisation and the adoption of a mode of liberal productivism would also involve a coherent programme of continued state paternalism and communal well-being. As one architect of the policy of securing trade liberalisation and economic revitalisation through the agency of foreign direct investment argued:

A modern community is concerned with collective as well as private spending; with the structure of education and its adequacy in relation to the world of tomorrow, and with the provision made for other social needs such as housing, health, social welfare and communications.

(cited in McCarthy 1990: 38)

The notion that economic nationalism might be replaced by an even more devoted welfarist approach, in which the drive towards industrialisation would not forsake the welfare of the weak, implied that the Republic's engagement with international capitalism could be guided by a consideration for the material security of every social class (Lee 1989). Clearly, the proponents of socio-economic rejuvenation found it convenient to explain and justify their modernist discourse by continuing the allegory of a collective past being reproduced in a less than traditional future. This fabricated relationship with the past was not, however, forthcoming in a future in which trade and social class liberalisation were to herald extended social inequality, endemic unemployment and the virtual destruction of the small farmer class synonymous with the traditional ideological representation of Ireland.

By the late 1950s, Fianna Fáil had rejected the framework of self-sufficiency and was striving to adopt a mode of development in which social objectives and cross-class unity might be achieved through state regulation, welfare provision and the neo-liberalisation of the labour market. In terms of class structure, the Republic was shifting from agriculturalism and family property towards one based upon urbanisation, industrialisation, skill and educational qualifications (Whelan 1995). More important, consumption and its administration was becoming a dominant mode of social relations. All of this was achieved through the 1960s and 1970s by virtue of pseudo-cosmopolitanism, a modernist discourse and an almost total submission to international capital (Smyth 1991).

Unsurprisingly, as in the previous era of economic nationalism, the socio-economic objectives of communality were not accomplished. If anything, the Republic now displays some of the worst extremes of social marginalisation within the EU. Its unemployment rate (21.8 per cent in 1995) is at a record high, being twice the average EU rate. Furthermore, the percentage of those employed within the total population fell from 35.7 per cent in 1970 to 30.1 per cent in 1995. During the period 1971–91 there was also an 8 per cent fall in active labour force participation and a 27 per cent growth in aggregate unemployment (OECD 1996; Shirlow 1995). The structural severity of unemployment has also been reflected in the extent and composition of long-term unemployment. In comparison with other EU states, the Republic has witnessed the highest growth in long-term unemployment from a below-average 36.7 per cent in 1983 to the third highest rate of 63.7 per cent in 1994 (Shirlow 1996). It is clear that, in terms of social class alteration, the main effect of transnationalism and the emergence of a post-agricultural mode of accumulation has been the generation of widespread income and class differentials, particularly between the employed and unemployed. The socio-economic position of the working classes and unemployed, who are over twenty times as likely to become long-term unemployed than their middle-class counterparts, has been characterised by expanded social marginalisation, attributable to the linked growth of poverty, unemployment, emigration and an inequitable distribution of social opportunity.

New class relationships in the Republic

Since the demise of economic nationalism, the subsequent growth in labour market inactivity during the period of economic restructuring has been accompanied by an enlargement in the socio-economic domination and material prosperity of the middle classes. In addition to large landowners and native industrialists, material ascendancy now encompasses elements of the administrative, technical and managerial professions. The revised class structures which have emerged have been heavily influenced by industrialisation, retailisation and financialisation, these rising sectors of production providing new opportunities for capital accumulation and intensified rates of technological and organisational innovation. More important, the middle class has managed to preserve and consolidate its material position – at the expense of the lower-income classes – through high levels of self-recruitment and a near monopoly of access to higher education and therefore the professions. As Whelan argues (1995: 351): 'The nature of Irish industrialisation has been such that the degree of disadvantage suffered by the working classes has been greater than conventional . . . analysis would suggest.'

The emergence of labour markets in which well-paid and secure employment is increasingly tied to a maximisation of educational qualifications – also the case in Northern Ireland – indicates that access to higher education is a major structural factor contributing to social unevenness and the perpetuation of social inequality. In the Republic's foreign manufacturing sector, for example, which controls 51 per cent of manufacturing-based employment, graduates command 82 per cent of technical and managerial positions (Shirlow 1996). In 1993, only 11.2 per cent of graduates in the Republic came from skilled and unskilled manual or unemployed backgrounds, despite these social classes constituting 52.5 per cent of the total population (Shuttleworth and Shirlow 1996). The professional, administrative and technical classes also increased their share of total income by 25.7 per cent between 1971 and 1994, compared to a 16.8 per cent decline among the skilled and unskilled manual, small farmer and unemployed classes (Shirlow 1996). In terms of international income comparisons, Lee claims that: 'Even in *per capita* income we have fallen much further behind every other state, except Britain, that was ahead of us sixty years ago' (1986: 92). However, this misses the essential point, for the Republic has the fourth highest rate of disposable income in the EU (Shirlow 1995). The key issue in terms of income distribution is that the lower-income groups receive an insignificant and disproportionate share of wealth creation.

This failure to produce a socio-economic regime capable of sustaining employment and domestic income growth among the lower-income classes has also been linked to the continuation and extension of class disparities and the perpetuation of a highly conservative political structure, largely controlled by property interests (Smyth 1991). The business-owning class is heavily subsidised through low corporation tax, massive state subsidies and relatively low levels of personal taxation. In addition to tax breaks, a range of state-sponsored agencies are almost entirely devoted to sustaining the private sector. This contrasts starkly with a working class negatively affected by austere reductions in welfare provision and other national debt-management policies (Porter and O'Hearn 1995). The bourgeoisie has emerged as a politically dependent, capitalist-based, property-owning class whose activities are largely predicated upon the exploitation of its relationship with Fianna Fáil and, to a lesser extent, Fine Gael (Keating and Desmond 1993). As Kirby remarks:

> It is precisely from these sections of small native capital that the day-to-day managers of the political system in the South are drawn. Parties like Fianna Fáil, at rank and file level as well as leadership level, are full of these small-time capitalists and local members of the middle classes.

> (Kirby 1988: 37)

Therefore, a contemporary social dualism has been created in the Republic by the declining social coherence of rural and urban lower-income localities – damaged by high levels of out-migration and increases in unemployment and underemployment – and the conspicuous consumption and relative career invulnerability evident among a wealthy local élite

(Kirby 1988). The argument advanced by commentators such as Lee (1989), Hazelkorn and Patterson (1995) and Breathnach (1995) is that such repercussions of social marginalisation reflect an absence of planning and ability at the core of an undynamic political system. The middle classes have been ineffective in articulating a socially responsible and indigenous-led economic strategy. Breathnach (1995: 24) accuses them of being 'imbued with a "quick buck" mentality', while Lee (1989: 612) argues that 'the causes of Irish retardation' are due to 'a failure to mobilise the intellectual resources of the country properly'. Although such observations are undoubtedly convincing, even they may underestimate the importance of contemporary class forces in the Republic. More robustly, the middle classes have been spectacularly successful, not only because they have guaranteed and extended their control over consumption, but because they have done so without unleashing class hostilities, an outcome explicable only through the depoliticising effects of mass emigration and the sense of social fatalism that this engenders among the lower-income classes.

Furthermore, these commentators do not address the issue that the middle classes, in reproducing their relative material prosperity, have abdicated national responsibility in favour of an almost total subservience to the ethos of liberal productivism and economic transnationalisation. The rejection of national accountability, indigenous-led development and the wholehearted adoption of monetarist and other neo-conventional strategies indicates that the new élite has opted to ignore the previous pretence of an allegoric devotion to – or responsibility for – other social classes. The result is a social unevenness that, over the past four decades, has produced a society increasingly characterised by drug addiction, organised crime and social retardation. Tradition has become little more than a commodity to be utilised in an increasingly capitalised form of tourism and consumption.

NORTHERN IRELAND: THE SOCIO-SECTARIAN CONTEXT

The establishment of Northern Ireland in 1921 was achieved through a careful territorial delineation which ensured a unionist majority. In order to perpetuate this hegemony, socio-spatial and class formations were constructed through distinctly sectarian practices. In terms of traditional class alignments, the ownership and administration of industry was predominantly Protestant-controlled. Similarly, craft-based and skilled employment was a virtual preserve of Protestant working-class males. The exclusion of Catholics from significant and relatively well paid areas of employment was demonstrated by their marked over-representation in the unskilled and female sections of the labour market and by higher levels of unemployment (Shirlow and McGovern 1996). Thus the socio-spatial realities of sectarian ascendancy meant that the uneven class distribution of wealth was accentuated by religious affiliation. However, such domination was not based solely on evident religious antagonism and socio-economic divisions, but was also conditioned by the anxiety of the unionist establishment that Protestant workers could be tempted by socialist/labourist politics or anti-sectarian discourses. Combined with the Catholic minority's support for reunification, such an alliance would have endangered the very survival of what had become a bourgeois-dominated unionist state (Bew *et al.* 1979).

In order to accommodate these actual, potential or perceived tensions, it was vital that the unionist élite fabricated a sense of socio-economic devotion towards Protestant workers in order to eliminate the possibility of the latter engaging in political activities beyond their control. The state's affirmation of sectarian structures remained an important element in the reproduction of what became identified as traditional cross-class relationships within the Protestant community. The primary basis of this agenda was the state's sensitivity to the political impact on the Protestant masses of any reduction in their material well-being relative to their Catholic counterparts. Thus unionist hegemony was reproduced through an economic and civic structure, which promoted and sustained, whenever possible, the comparative material interests of working-class Protestants. This was achieved by ensuring that post-1945 welfare policies were applied in Northern Ireland and by actively promoting discriminatory employment practices. For example, up until the 1960s, many job advertisements stated explicitly that 'Protestants only need apply'. Furthermore, voting and welfare allocation structures were often tied to maintaining unionist authority, shown, for example, by the notorious gerrymandering in Derry City (McCann 1993). This paternalised mode of sectarianism purposely exacerbated socio-religious division and also earned the loyalty of Protestant workers towards the state and employers (McGovern and Shirlow 1996).

By the 1960s, however, the uneven nature of political representation and material conditions combined with the onset of de-industrialisation to challenge the unionist hegemony. On one hand, the demands for social democracy and ideas of social citizenship – which emerged throughout western society during the 1960s – influenced a Catholic minority, which began collectively to oppose a social system that undermined its civil and constitutional rights. Simultaneously, though, Protestant workers were deeply affected by the continuing loss of jobs in traditional industries, which formed the core of their employment opportunities and provided access to comparative material well-being and privilege. De-industrialisation also threatened a hegemonic unionist bourgeoisie whose assertion of political authority was founded on a paternalistic relationship with working-class Protestants.

Thus the administration of social welfare and industrial location became crucial points of contention as both Catholics and Protestants looked for signs, respectively, of transformation and continuity. Unable to break away directly from sectarianised practices, the Unionist Party continued to support the location of capital and welfare investments in majority Protestant districts, largely at the expense of the predominantly Catholic and socially impoverished western part of Northern Ireland (Purdie 1991). The perpetuation of such manifestly discriminatory practices destroyed any Catholic expectations of a non-sectarian future. However, their demands for the cessation of such policies fanned the fears of many Protestants, who deciphered such claims as a challenge to their 'rights' to state-directed paternalism.

The Unionist Party was caught in a distinct socio-cultural quandary. It could not covenant a new collective future as this would mean responding positively to the Catholic minority's grievances by enforcing non-sectarian practices, which would ultimately disrupt unionist solidarity and identity. Furthermore, unlike its Fianna Fáil counterparts in the Republic, the unionist establishment had never articulated a collective populist discourse in which there was any notion of a consolidated Northern Irishness intended to include both religions and all social classes (see Chapter 10). Instead, it had enforced a form of sectarian populism and a mode of socio-economic integration which visibly excluded a Catholic population whose faithfulness it had never sought. By the early 1970s, the unionist establishment had thus become imprisoned within the constraints and contradictions of its own form of sectarianised populism, unable to react to a new socio-economic order, in which employers maximised educational qualifications and promoted social mobility, because it would break the connections between access to employment and relative Protestant well-being. The eventual results were to be civil war and British state intervention.

In terms of religious affiliation, the eventual dissolution of traditional class alignments and unionist hegemony was experienced in very divergent ways. For many Protestant workers, the failure of their state to regulate and ultimately overpower Catholic-led hostility and socio-political demands fuelled the re-emergence of loyalist paramilitary groups. Their activities, allied to doubts concerning the impartiality of policing and the failure of the state to respond adequately to the socio-economic demands of Catholics by dismantling traditional sectarian policies, encouraged the remobilisation of the Irish Republican Army and the use of violence in order to remove what was perceived as a state beyond reform. In the violent crisis, stimulated by the collapse in the regulatory capability of the Unionist government, the British state imposed Direct Rule in 1972 and so began the slow process of rebuilding a new form of socio-political control.

Paralleling events in the Republic, this epoch of state intervention has been defined by the need to counterbalance market failures and promote alternative economic growth structures (Murtagh 1993). The socialisation of labour in relation to the conditions of capitalist production in Northern Ireland, as elsewhere, has been constructed around various mechanisms of social control, persuasion, education, training and the

mobilisation of new social forces. However, Northern Ireland is exceptional in that the mobilisation of social forces has been continually tied
to, and dictated by, the unleashing of distinct sectarian sentiments. In
other words, socio-economic conditions are continuously understood in
relation rather to competing sectarian and class identities than to the
resolution of accumulatory crises and the fabrication of new modes of
economic activity.

Regulating social forces in Northern Ireland

Since the inception of Direct Rule, the onus placed upon the British state
has been to develop a series of socio-regulatory practices which might limit
the challenge to its overall legitimacy (Tomlinson 1993). This aim has been
pursued through a policy of socio-political normalisation and the adoption
of practices whose primary goal is to secure the construction of a set of social
relationships which, it is hoped, will transcend sectarian hostilities and
engender socio-economic normality (Shirlow and McGovern 1996; Smyth
and Cebulla 1995). The state has also sought to police and/or contain the
conflict as well as accommodate new socio-economic strategies. Economic
activity and labour markets have been reorganised to facilitate cross-border
co-operation and the absorption of middle-class Catholics into a new political consensus. However, the erosion of traditional class alignments and the
attempted de-sectarianisation of civil society have produced limited as well
as contradictory results, since the unleashing of new social forces has both
blunted and reproduced sectarianism.

While sectarian animosity is still visible among all social classes, a
growing body of evidence supports the thesis that the middle classes, irrespective of their religious affiliations, increasingly share similar lifestyles and
socio-economic pursuits, which are mutually agreeable and inherently less
antagonistic. The emergence of a sizeable Catholic middle class is indicative
of social mobility, but may also attest to a form of socio-cultural realignment. In turn, middle-class Protestants are now more likely to embrace – or
at least tolerate – various all-Ireland institutions. The business community
in particular is anxious to stimulate cross-border trade, company mergers
and the sharing of industrial know-how. Finally, the middle classes are more
likely to socialise with, and marry, members of the opposite religion than is
the case in the lower-income groups (Felderly 1994). The latter have
emerged from twenty-five years of conflict more divided than ever in terms
of residence, cultural affiliation and political identity. Sectarian assassination, rioting and conflicts over marching point to a socio-economic group
essentially involved in intra-class conflict. While this does not mean that it
is impossible to locate non-sectarian and shared identities among many
members of the working classes, it is more difficult to identify such attitudes than among the middle classes (O'Toole 1994).

In seeking to address the problems created during the previous epoch of unionist hegemony, the British state has endeavoured to reconstitute middle-class political affiliations by reformulating the nature of competing political identities. To this end, socio-economic policy-making and anti-sectarian legislation have been implicated in an attempt to forge a third tradition, capable of living with evident cultural and political ambiguities (O'Connor 1993; O'Toole 1994). This could include those who favour, for example, integrated schooling and who are prepared to share in a more pluralistic political programme, perhaps countenancing joint-sovereignty or a form of power-sharing. This third tradition also accepts the logic of a degree of cross-border reconciliation, which might both deliver such political goals and help develop and enlarge the productive capacities and market opportunities of an all-Ireland economy.

The emergence of a post-industrial economy, flexible labour markets, de-industrialisation and extended income inequality have thus been guided by both commonplace socio-economic shifts and also the purposeful manipulation of the class structure in order to alter the socio-political fabric of Northern Ireland. In terms of the province's social structure, two broad trends can be discerned. First, white-collar employment has grown because, throughout the conflict, the British state has ensured the reproduction of sufficient outlets for professional and administrative employment and investment opportunities, even though such a policy is largely unprofitable and stands in stark contradiction to the monetarist mode of accumulation evident throughout the rest of the United Kingdom (Gaffikin and Morrissey 1990). Second, there has been a rise in unemployment, low-paid employment and underemployment among the increasingly marginalised Protestant and Catholic working classes (McAuley 1994; Smyth and Cebulla 1995). The British state has not attached the same dynamism or commitment to creating intra-class solidarity among those on low incomes, reflecting a policy and form of containment based upon spatial and ideological considerations. The spatial policy has been to restrict the conflict, whenever possible, to certain low-income areas. Ideologically, the problem was to be contained among those who, it was erroneously perceived, could not transcend inter-communal conflict, the people allegedly referred to by one previous Secretary of State as the 'brutal and murderous Ulster working-classes' (Felderly 1994: 32).

As the state has arguably striven to protect and promote the position of the middle classes, it has also done less to protect traditional labour market structures. Their erosion has impacted heavily upon the material well-being of the working classes, for whom labour-market conditions, particularly between 1971 and 1995, have become extremely inhospitable. A 42.7 per cent growth in non-earner families, a three-fold increase in poverty, a 32.2 per cent decline in income, and a rise in unemployment from 4.3 per cent to 14.2 per cent all attest to social dislocation (Borooah 1993; Shuttleworth and Shirlow 1996; Teague 1993). Conversely, the evolution of an economy dominated by the private and public service sectors has significantly bene-fited the middle classes. Since 1971, they have enjoyed a 28.1 per cent rise in their share of total income and now possess the highest levels of personal savings as a percentage of disposable income within the United Kingdom (Felderly 1994).

This process of maintaining and enlarging the socio-economic dominance of the middle classes by promoting a service-led economy has tended to create an unequal profile in terms of class – as opposed to religious affiliation – within certain sections of the labour market. Nearly a third of the jobs created between 1990 and 1995 were professional, managerial or adminis-trative positions. Although the Catholic share of such employment has grown by 32.8 per cent since the mid-1970s, over 70 per cent of successful applicants for such positions had third-level education and came from the middle classes (data supplied by Fair Employment Commission). Thus, the benefits paradoxically accruing from twenty-five years of conflict are denied to many working-class Catholics, ghettoised into low-paid employment and unemployment (O'Connor 1993; Shirlow and McGovern 1996). Equally, many working-class Protestants are denied entry into the new labour markets, requiring educational qualifications or skills that they do not possess. However, this is rarely acknowledged as the consequence of economic restructuring and uneven class forces. Instead, explanations are couched in terms of abandonment by middle-class Protestants and a rising socio-economic ascendancy of the Catholic population. In the often displeasing and hopeless space created by these duplicated processes of alien-ation and social marginalisation, the nature of contemporary Loyalism and Republicanism has been forged.

As a result, working-class Catholics and Protestants are experiencing increasingly similar degrees of social marginalisation. Although Catholic males are twice as likely to be unemployed as their Protestant counterparts, this masks a doubling of Protestant male unemployment between 1971 and 1995. In terms of average household incomes within the Belfast Urban Area (BUA), the percentage share of those whose average incomes are 50–80 per cent less than the Northern Ireland average are relatively similar for Protestants (63.6 per cent) and Catholics (69.6 per cent). In 1971, the six most deprived Protestant wards in the BUA had household incomes which were 18 per cent higher than in the six most deprived Catholic wards. By 1991, the gap had narrowed to 7.2 per cent (McGovern and Shirlow 1996).

The erroneous perception held by sections of the Protestant community that its economic plight is due to British state-inspired appeasement of Sinn Féin voters and a subsequent growth in employment for working-class Catholics has resulted in the sectarian targeting and murder of Catholics. Such events represent the most obvious and extreme response to the notion of a Catholic-inspired loss of traditional Protestant status. Similarly the imagined relationship between Protestant workers and the state is still constructed around the notion that Protestant faithfulness should be rewarded economically, precisely the same allegoric relationship which underwrote the previous mode of sectarian domination and unionist hegemony.

For lower-income Catholics, the abundant evidence of their poor socio-economic status clearly influences support for Sinn Féin and the politics of reunification (Bean 1996). Since partition, the subordinate status of this class, and its inability to achieve widespread social mobility, has inspired the notion of a community ghettoised and abused by the economic dominance of the Protestant community and the political activities of the British state. Given this mind-set, it is unsurprising that as social conditions deteriorate, republicans view Northern Ireland as little more than a sectarian statelet. In reality, however, they are experiencing a form of social discrimination also shared by the Protestant working class. Both groups are condemned by a lack of the skills or talents now required by the labour market, especially in relation to securing well-paid employment.

Nevertheless, it is also clear that discrimination, whether social or religious, impacts more heavily upon lower-income Catholics than it does upon the Catholic middle class, a situation which has led McCann to comment that 'in business, commerce and the professions, there's no disadvantage in being a Catholic' (1993: 52). In effect, middle- and working-class Catholics are involved in different labour and vacancy markets and, moreover, experience the realities of sectarianised space in different ways. For the middle classes, both Catholic and Protestant, the office or business arena is less sectarian and as a result relatively secure. Conversely, for the working classes, the journey to work can mean passing through areas dominated by the other religion or working on a shop-floor in which religious or cultural identity can lead, as it has on many occasions, to the symbolic placing of bullets in lunch boxes, verbal abuse and, worse, physical assault and assassination.

One survey has shown that the prevalence of 'chill factors' (the perception that security is not guaranteed) has created a situation in which 60 per cent of lower-income Protestants and Catholics would not work in a place 'predominantly of the other religion'. Catholics in west Belfast are twice as likely to seek work in Germany as compared to east Belfast, while 73 per cent of Protestant males residing in the eastern Waterside area of Derry City would not accept work on the predominantly Catholic west bank (Shuttleworth and Shirlow 1996). The combination of such factors with the continuing existence of discrimination and the realisation that private-sector employers have not actively striven to achieve balanced work-forces, clearly influences working-class Catholics, who cannot owe allegiance to a form of fair employment legislation that has ultimately failed to respond to their obvious grievances.

Unlike their Protestant counterparts, working-class Catholics comprehend their poor socio-economic position not as the product of betrayal but as perpetuated abuse. As such, they are not reconciled with – or prepared to accept – the Northern Ireland state because they remain, as previously, visibly excluded from engaging in social mobility and/or employment. It is not surprising, therefore, that Sinn Féin, which attracts a third of its support from unemployed Catholics, continually seeks to expose the substantial residue of religious discrimination (McGarry and O'Leary 1995). Obviously, the motives behind voting for Sinn Féin are not linked to direct materialist factors alone – they also include protesting against secondary citizenship and rebuking more conformist middle-class Catholics. Thus the emergence of new class forces is impacting upon the cohesion of a Catholic population fragmented by differing material experiences and degrees of *rapprochement* with the British state. As O'Connor notes (1993: 18), 'Not surprisingly, the perception that British direct rulers have none of the discriminatory instincts of Unionist governments is more enthusiastically expressed among Catholic civil servants and lawyers than it is in unemployment blackspots like Strabane.'

The material prosperity of middle-class Catholics underscores the importance of access to education and anti-discriminatory practices in relation to opening new avenues of social mobility. However, the political attitudes and national identity of this class are more difficult to determine. There is no denying that a majority of its members support some form of reunification, but the manner in which this is to be achieved, and what it would actually represent, is far from clear. Surveys (conducted by the author) among middle-class Catholics indicate that almost 50 per cent support a negotiated settlement, which would produce a reunited Ireland. Virtually as many, however, claimed that they would accept joint-authority, either as a primary or secondary option. Within this latter group, almost two-thirds agreed with the statement that: 'The two parts of the island are so different that shared sovereignty and state reform as opposed to full-scale reunification would be more suitable.' This probably reflects the fear, much promoted by unionists, that reunification would be economically disastrous. Indeed, 20 per cent of the sample upheld the union perspective and accepted that power-sharing would be the most satisfactory outcome. A similar percentage supported Sinn Féin, agreeing with the statement that: 'British withdrawal and one-step unification is the only solution.'

Such a disparate range of political positions suggests that the Catholic middle class possesses neither a unitary notion of national identity nor a distinctive sense of cultural allegiance to Irishness. This fragmentation of Catholic identity has been conditioned by two forces. First, social mobility has obviously consolidated material well-being as well as diluting the attractiveness of a reunited Ireland. Second, republican violence and the growing perception that the Republic is relatively agnostic about reunification has created a significant group of Catholics who are either pro-union or quasi-pro-union. Such opinions and perceptions sit in stark contrast to those members of the low-income Catholic electorate who habitually support Sinn Féin.

Middle-class Protestants have also been influenced by the onset of Direct Rule as they are no longer the sole embodiment of political and socio-economic power in Northern Ireland. In addition to its deprivation of political authority and power, this class is now less likely directly to control or dominate industries with large work-forces. It cannot therefore publicly promote sectarian practices or reproduce its hegemony by immediately influencing the activities of Protestant workers. Middle-class Protestants have, as a result, become a subsidised class which has shifted away from an Ulster towards a British identity (McGarry and O'Leary 1995). In so doing, they have integrated themselves fully into the institutions of Direct Rule. Furthermore, middle-class Protestants have withdrawn from direct political representation to such an extent that the Ulster Unionist Party is now openly engaged in luring them back into the party political affairs of Northern Ireland.

However, the reality of British state-enforced autocracy, following the failure of the unionist hegemony to regulate civil society, has effectively eroded the principal reasons for Protestant middle-class participation in party political life. Increasingly, the more prosperous members of this class tend to mobilise their collective energies and influences through civil society and professional bodies. As is also true of their Catholic counterparts, they operate in a dissimilar way to the Republic's middle class, which openly mobilises political influence and a mode of clientelism to advance its material position. In effect, the middle classes in Northern Ireland, both Protestant and Catholic, tend to operate as an arm of the British state – and increasingly the EU – in terms of gaining and securing extended material prosperity. The result is a less politicised and combined class which has exchanged direct political control in favour of a more paternalised relationship with Westminster and, to a lesser extent, Brussels.

Thus in terms of traditional class alignments, Northern Ireland has shifted from a hegemonic and industrial form of state-regulated sectarianism towards a post-industrial structure. This shift has been manipulated by a British state ensuring social mobility for middle-class Catholics and also pushing their Protestant counterparts away from their previously hegemonic past. The central aim is to tie the middle classes so tightly to the structures of the British subvention that they cannot operate or reproduce their material well-being without recourse to the British state. The expectation is that they will ultimately owe allegiance to the subvention, which underwrites their material position, and in turn dispense with traditional socio-religious hostilities. For the lower-income classes, this manipulation of resources has not improved their socio-economic profile as they have been socially alienated due to extended income inequality. It is not surprising therefore that traditional sectarian hostilities are now visibly located and replayed among a lower-income class embroiled in an intra-class war. The potency and significance of this conflict removes the adoption of an alternative political discourse of solidarity among the working classes.

CONCLUSION

Although both parts of Ireland now display superficially similar socio-economic profiles in terms of class and material composition, the two economies have gone through somewhat dissimilar processes of socio-economic and cultural transformation. Furthermore, due to socio-religious conflict, the mitigation of Northern Ireland's viability as a distinct economic entity and the ensuing economic dependence upon the state sector, the nature of socio-economic transformation in Northern Ireland has evidently been more politicised than is the case in the Republic. Here the loss of a coherent national programme and the adoption of pseudo-cosmopolitanism with its concomitant social marginalisation underline the break with any previous notion of devoted nationalism. The pragmatic way in which Irish society shifted from this discourse to fully fledged modernism may well indicate the fragility of Irish identity and the triumph of materialism over communal well-being (Kirby 1988). Moreover, the inability to mobilise a contemporary form of 'communal devotion' in order to tackle social inequality means that there no longer remains any cross-class collective experiences or bonds of solidarity that might produce a more stable and equitable society.

The result of class transformation throughout Ireland has been the production of modes of accumulation which have hindered the elaboration of a more equitable social structure. In the Republic, the notion of a nationally devoted economy has been replaced by the free-fall of trade liberalisation and the resultant material retardation experienced by the lower-income classes. In terms of a collective identity, it is now evident that the middle classes, North and South, increasingly possess a transnationalised identity in which relationships with London, Brussels, Washington and Tokyo predominate over a previously strong association with their respective parts of Ireland. In the Republic, in particular, the primacy of national life and cultural coherency, long the embodiment of the national ideal, has been cast aside as the middle class has lost faith in the previous discourse tied to populist socio-economic policies.

What we are left with on the island of Ireland is a more unified class structure attributable to the conditions of post-industrialism. Nevertheless, the actuality of socio-economic reproduction and the political economy of both countries is tending towards divergence. For example, the Republic's economy is firmly linked to the attraction of mobile capital investment, political stability and external subsidisation. Conversely, in Northern Ireland socio-economic conditions are shaped by conflict and, whenever possible, the manipulation of social forces by the British state. More crucially, the working-class communities in Northern Ireland are drawn ever more closely into intra-class conflict. It is not surprising, therefore, that the Irish government is becoming relatively agnostic about reunification and the possibility of having to regulate the socio-sectarian conflict that is constantly being reproduced in Northern Ireland. The development of productive and social forces and the dissolution of traditional class alignments have been tied to a clear demarcation and redefinition of class interests and opportunities. Both the adoption of the contemporary process of economic development and its failure to challenge the perpetuation of social heterogeneity divide and impede the development of a more equitable Irish society. The political systems, in both Northern Ireland and the Republic, and their dedication to middle-class interests, have impeded the evolution of coherent and systematic alternatives, capable of tackling the nature of social dislocation and socially destabilising social forces.

REFERENCES

Bean, K. (1996) 'The new departure', *Causeway* 6: 202–13.

Bew, P., Gibbon, P. and Patterson, H. (1979) *The State in Northern Ireland*, Manchester: Manchester University Press.

Borooah, V. K. (1993) 'Northern Ireland – typology of a regional economy', in P. Teague (ed.) *The Economy of Northern Ireland: Perspectives for Structural Change*, London: Lawrence and Wishart.

Breathnach, P. (1985) 'Rural industrialisation in the West of Ireland', in M. J. Healey and B. W. Ilbury (eds) *The Industrialisation of the Countryside*, Norwich: Short Run Press.

—— (1995) 'Uneven development and Irish peripheralisation', in P. Shirlow (ed.) *Development Ireland: Contemporary Issues*, London: Pluto Press.

Felderly, E. (1994) *'Fight For It. Or Get Out'*, Ontario: Conflict Press.

Foley, A. and McAleese, D. (eds) (1991) *Overseas Industry in Ireland*, Dublin: Gill and Macmillan.

Gaffikin, F. and Morrissey, M. (1990) *Northern Ireland: The Thatcher Years*, London: Zed Books.

Gibbons, L. (1996) *Transformation in Irish Culture*, Cork: Cork University Press/Field Day.

Hazelkorn, E. and Patterson, H. (1995) 'The new politics of the Irish Republic', *New Left Review* 211: 49–71.

Keating, P. and Desmond, D. (1993) *Culture and Capitalism in Ireland*, Aldershot: Avebury.

Kirby, P. (1988) *Has Ireland a Future?*, Cork: Mercier.

Lee, J. J. (1986) 'Whither Ireland? The next twenty-five years', in K. Kennedy (ed.) *Ireland in Transition: Economic and Social Change Since 1960*, Cork: Mercier.

— (1989) *Ireland 1912–1985: Politics and Society*, Cambridge: Cambridge University Press.

McAuley, J. (1994) *The Politics of Identity*, Aldershot: Avebury.

McCann, E. (1993) *War and an Irish Town*, 3rd ed., London: Pluto Press.

McCarthy, C. (1990) 'Outline for the 1990s – a decade of fiscal restraint?' in D. Kennedy and C. McCarthy (eds) *Prosperity and Policy: Ireland in the 1990s*, Dublin: Institute of Public Administration.

McGarry, J and O'Leary, B. (1995) *Explaining Northern Ireland*, Oxford: Blackwell.

McGovern, M. and Shirlow, P. (1996) 'Sectarianism, regulation and the Northern Ireland conflict', *Reclus: Journal de L'Espace Geographique* 37: 13–32.

Murtagh, B. (1993) *Planning and Ethnic Space in Belfast*, Coleraine: University of Ulster Press.

O'Connor, F. (1993) *In Search of a State: Catholics in Northern Ireland*, Belfast: Blackstaff.

OECD (1996) *International Investment and Multinational Enterprises*, Paris: OECD.

O'Hearn, D. (1993) 'Global competition, Europe and Irish peripherality', *The Economic and Social Review* 24, 2: 169–97.

O'Toole, F. (1994) 'Floating unity on a tide of people', *The Guardian* 23 February 1994.

Peet, R. (1991) *Global Capitalism*, London: Routledge.

Porter, S. and O'Hearn, D. (1995) 'New Left "rodsnappery": the British Left and Ireland', *New Left Review* 212: 66–86.

Purdie, B. (1991) 'The Demolition Squad', in S. Hutton and P. Stewart (eds) *Ireland's Histories: Aspects of State, Society and Ideology*, London: Routledge.

Shirlow, P. (1995) 'Contemporary development issues in Ireland', in P. Shirlow (ed.) *Development Ireland: Contemporary Issues*, London: Pluto Press.

—— (1996) 'Transnational corporations in the Republic of Ireland and the illusion of economic well-being', *Regional Studies* 29, 7: 687–705.

Shirlow, P. and McGovern, M (1996) 'Sectarianism, socio-economic competition and the political economy of Ulster Loyalism', *Antipode* 82: 123–47.

Shuttleworth, I. and Shirlow, P. (1996) 'Vacancies, access to employment and the unemployed', in E. McLaughlin (ed.) *Policy Aspects of Employment Equality*, Belfast: SACHR.

Smyth, J. (1991) 'Industrial development and the unmaking of the Irish working class', in S. Hutton and P. Stewart (eds) *Ireland's Histories: Aspects of State, Society and Ideology*, London: Routledge.

Smyth, J. and Cebulla, A. (1995) 'Industrial collapse and the post-Fordist overdetermination of Belfast', in P. Shirlow (ed.) *Development Ireland: Contemporary Issues*, London: Pluto Press.

Teague, P. (1993) 'Discrimination and fair employment in Northern Ireland', in P. Teague (ed.) *The Economy of Northern Ireland: Perspectives for Structural Change*, London: Lawrence and Wishart.

Tomlinson, M. (1993) 'Policing the New Europe: the Northern Ireland factor', in T. Bunyan (ed.) *Statewatching in the New Europe: A Handbook on the European State*, Nottingham: Statewatch.

Whelan, C. (1995) 'Class transformation and social mobility in the Republic of Ireland', in P. Clancy, S. Drudy, K. Lynch and L. O'Dowd (eds) *Irish Society: Sociological Perspectives*, Dublin: Institute of Public Administration.

Wickham, J. (1985) 'Dependence and state structure: foreign firms and industrial policy in the Republic of Ireland', in O. Holl (ed.) *Small States in Europe and Dependence*, Vienna: Braumuller.

6

EMBODIED IRISHNESS

Gender, sexuality and Irish identities

Catherine Nash

INTRODUCTION

Ireland has a long history of being represented as feminine. In turn, versions of Irish national identity have prescribed certain kinds of gender and sexual identities for Irish men and women. These gendered representations of Ireland and Irish gender identities impact upon the lives of women and men and influence their opportunities and constraints in work, education, political activity, personal relationships and senses of themselves. This double sense of embodied nationhood forms the focus of this chapter, which investigates the relationship between the representation of Ireland as female and the construction of gender and sexual identities in Ireland. Irish feminist activists, artists and writers have pointed to the ways in which the gendering of Ireland and national versions of ideal Irish femininity have been deeply problematic and damaging for women. Feminist work on the relationship between national, gender and sexual identities has explored the ways in which they are mutually constructed in geographically, historically and culturally specific manners. Because the rights and welfare of individuals and definition of citizenship in the nation-state are differentiated according to gender and sexuality, concepts of nationhood and national identity have been criticised but also reworked to avoid traditional patterns of exclusions and exclusiveness. Two examples can introduce some of the issues involved in the conjunction of the historical and contemporary imagining of the nation and the construction and experience of gendered and sexual identity. Both constitute interventions into the meaning of Irishness.

Firstly, much of Eavan Boland's poetry has addressed her sense of the dissonance in Ireland between being a women and being a poet within Irish national poetic traditions (Boland 1989, 1996). Her attempts to write of her own experience and the experience of women in the past in Ireland has involved interrogating the meaning of nationhood and history and reformulating a poetic tradition in which women have figured as passive symbols of the nation but which has largely ignored the experiences of women. She writes:

> I thought it vital that women poets such as myself should establish a discourse with the idea of the nation. I felt sure that the most effective way to do this was by subverting the previous terms of that discourse. Rather than accept the nation as it appeared in Irish poetry, with its queens and muses, I felt the time had come to re-work those images by exploring the emblematic relation between my own feminine experience and a national past.
>
> (Boland 1989: 20)

The second example is the case in which, after a series of previous injunctions, the organisers of the 1994 Boston St Patrick's Day Parade cancelled the event rather than allow the Boston Irish Gay, Lesbian and Bisexual Group to take part. Kathleen Finn, a spokesperson for the group, expressed her sense that it was 'being put on trial for what it means to be Irish', since for the organisers 'somehow being heterosexual is so wrapped up with being Irish that they simply can't imagine someone being Irish and not being heterosexual' (Finn 1995: 7).

A number of initial points can be made here. Firstly, the different location of these examples reflects the way in which ideas of Irish identity are produced not only by people living on the island of Ireland but are also articulated by groups and individuals in other places. The expression of immigrant senses of Irish identity may undermine the idea that there is only one true Irishness and that this depends on a stable and secure relationship to place. Reflections on leaving can also expose the cultural and social forces in Ireland which prompt moves away from contexts in which gender, sexuality and cultural identity are rigidly defined (Crone 1988; Smyth 1991). Yet as the case of the Boston parade indicates, diasporic versions of Irishness are not automatically inclusive. These emigrant versions of Ireland can return to influence ideas of identity in Ireland in radical and conservative ways. More broadly, there is a long history of definitions of Ireland and Irishness being constructed through often unequal encounters between different groups and cultures in the island and elsewhere. Secondly, ideas of Irish national identity are made and communicated through cultural forms – for example, through a poetic tradition and parade, both with their particular stylistic and symbolic elements. Thirdly, cultural expressions or enactments of identity are often moments of conflict – here between traditional ideas of gender identity and sexuality in Irish national traditions and attempts to combine an attachment to Irish cultural forms with alternative expressions of femininity or sexuality.

Within most national traditions individuals are assigned certain kinds of sexual and gender roles and identities. These ideas of appropriate roles and identities for men and women have material effects. While they obviously impact on women and men differently, they are also experienced in different ways according to class, age, sexuality and geographical location. For example, the poet Nuala Ní Dhomhnaill (1994: 171) describes how, in the late 1950s, her parents returned with her from England to Ireland. Both her parents were doctors, yet on returning her mother could not practice her profession as she had done in England and as her husband continued to do, because of legal restrictions in Ireland at that time on married women working. Legislation regarding the family, divorce, employment and reproductive control in Southern Ireland this century, and the marginalisation of issues of gender, sexual and economic inequality in favour of the 'cause' for or against British involvement in Ireland, are very much bound up with the ways in which gender and nation are understood and represented. Cultural constructions of nationhood, gender and belonging are inseparable from the organisation of society and nature of politics in Ireland in the present and the future. While Irish masculinities are as much constructed as femininities, and men have suffered as well as enjoyed the social roles and behaviours expected of them, the comparatively greater constraints and disadvantages that women experience reflect broader patterns of gender inequality.

The complex connections between gender and national identity in Ireland reflect a specific history of colonisation and trajectory of modernisation in Ireland. This history includes the early and sustained experience of British colonialism in its political, economic and cultural aspects, the changing class structure and land ownership pattern in post-Famine Ireland, the development of Irish cultural nationalism in the nineteenth century, the militarisation of Irish politics in the early twentieth century and the close links between religion and political power in both parts of Ireland after partition. Yet the feminisation of Ireland and the construction of Irish gender identities are also connected to broader Western frameworks of knowledge and understanding. Three points can be made here. First, within Western epistemology the world has been ordered through dualistic opposites, which define objects, people, places and qualities through ideas of absolute difference between positive and valued and negative and inferior categories and characteristics. As this chapter will discuss, both colonial and nationalist constructions of Irish identity have used contrasts between ideas of the civilised and savage, order and disorder, and between notions of masculinity and femininity. Second, images of women have been used in Western culture as symbols which stand for other concepts such as vulnerability, charity, chastity or corruption. In this way national identity has often been defined through gendering the country. Contrasts between cultures that are dubbed masculine or feminine are also invoked by confronting women of different cultures – for example, defining Irish identity by juxtaposing Irish and English women. Third, the gendering of Ireland and the construction of Irish femininity have been supported by the traditional associations between nature, land, fertility and femininity. The way in which these concepts have figured in the definition of nationhood and gender iden-tities will be an important theme in the sections which follow.

This discussion of gender, sexuality and Irish national identity does not attempt to provide an exhaustive account of their intersection. Instead, the chapter focuses on selected examples in order to show how attention to issues of gender and sexuality undermines the naturalness of normative versions of Irish masculinity and femininity and exposes the limits of national belonging. Women's reflections upon Irish nationhood and gender, in historical research, creative forms and critical writings within and outside Ireland, do not simply criticise dominant ideas of nationhood but offer new ways of thinking about Irish history, gender and cultural belonging. Thinking critically about nationhood, gender and sexuality is never simply about women but about understandings of Irish history, culture and identity. The substantial changes in attitudes and legislation that have occurred in the last two decades in the Republic and the efforts of women in Northern Ireland in campaigning for social justice and political change reflect the strength and diversity of the women's movement in Ireland. By engaging critically and creatively with Irish cultural traditions, women in different fields are attempting to reconceptualise ideas of the nation, national tradi-tions and relationships to place in more open and inclusive ways. Through the examples, I trace the gendered and geographical aspects of the concep-tual and material boundaries of the nation, and how they are bound up with ideas of the relationship between place, history, culture and belonging, the gendering of land and nature and gendered divisions between public and private space. In order to discuss these critical perspectives it is first neces-sary, however, to trace the diverse origins of the feminisation of Ireland and to consider how this representation of Ireland as female has been connected to the construction of masculinities and femininities and to unequal power relations between England and Ireland and between men and women.

FEMINISING IRELAND

The representation of Ireland as female has been used to define Ireland and Irishness in different ways, in different contexts and for different purposes. Those who have depicted Ireland as female have done so in order to make sense of, order and justify relationships between countries and claims to territory. In doing so they have drawn on and reinforced ideas of masculinity and femininity and certain kinds of relationships between men and women. Thus the gendering of Ireland has been used to define the cultural identity and political status of the society and the identities and roles of men and women in Ireland. While nationalist writers in the late nineteenth and early twentieth centuries drew on Irish mythological traditions, they did so in ways which were connected to the long and complex history of gendered and sexualised discourses of

Irishness, including colonial representations and interventions in Ireland and forms of resistance to them.

In the Gaelic traditions of pre-Christian Ireland, both the idea of sovereignty and the land of the kingdom was represented as a woman. This sovereignty goddess validated the right of the king to rule, and her condition and the condition of the land itself reflected the quality of the king who married her. Her appearance as old and ugly or young and beautiful was a measure of the king's political authority and merit. When the man was worthy, intercourse with her bestowed kingship on him and youth and beauty on her (Cullingford 1987: 4). This trope occurs also in medieval Irish mythology, where individual noble women reflect the condition of the king and kingdom through their relationships with men (Cairns and O'Brien Johnson 1991: 3). Though these accounts of the sovereignty goddess have been deployed in Irish cultural nationalism and, more recently, used to reclaim models of feminine power for women, the production of images of Mother Ireland do not simply derive from this tradition. The representation of Ireland as a woman is a particular example of the gendering of nations in colonial projects of subordination and national strategies of resistance. Inevitably, much of Irish nationalist discourse derives not from a deep unbroken tradition but from reactions to the experience of colonisation – including colonial attempts to fix the character of colonised subjects.

In the early modern period Ireland, like other colonies or potential colonies, was figured as female in ways which naturalised colonial penetration and regulation. As mysterious and unknown territory she must be explored and made known; as wanton woman she evokes disgust and must be tamed. In the carefully eroticised geography of the English colonial administrator, Luke Gernon, the country clearly invites male penetration and can only be truly tamed, ordered and made productive by male – and implicitly English – intervention.

> This Nymph of Ireland, is at all points like a young wench that hath the green sicknes for want of occupying. She is very fayre of visage, and hath a smooth skinn of tender grasse. Indeed she is somewhat freckled (as the Irish are) some parts darker than other. . . . Her breasts are round hillocks of milk yeelding grasse, and that so fertile, that they contend with the vallyes. And betwixt her legs (for Ireland is full of havens), she hath an open harbour, but not much frequented. . . . It is nowe since she was drawn out of the womb of rebellion about sixteen years, by'r lady nineteen, and yet she wants a husband, she is not embraced, she is not hedged and ditched, there is noo quickset putt into her.
>
> (From Luke Gernon 'A Discourse of Ireland' *c.* 1620,
> quoted in Hadfield and McVeagh 1994: 66)

Sixteenth- and seventeenth-century colonialists not only feminised Ireland but directed their attention to the characteristics of Irish women (Carroll 1993; Sharkey 1994). The apparent freedom and sexual promiscuity of Irish women was used as proof of the barbarity of the people. Irish women were also the focus of concerns about the maintenance of cultural difference and thus political power between the native Irish and New English settlers. Through the traditions of fostering children between families, intermarriage and wet nursing, Irish women could undermine the cultural purity of the English colonists, thereby destabilising the distinctions between barbarity and civility upon which political control was legitimated. Jones and Stallybrass (1992) show how an Act of 1537, which forbade the wearing of Irish cloaks known as mantles, was part of the symbolic and material politics of colonisation in the sixteenth and seventeenth centuries in which concerns about cultural difference, gender, sexuality and political authority were woven together. By then, rather than hoping to Anglicise the Irish, colonists insisted upon the absolute and hierarchical difference between themselves and the Irish.

In this context, the assimilation of the Old English descendants of the twelfth-century Anglo-Norman colonisation into Gaelic Irish culture, through adopting the mantle for example, threatened the stability of English cultural difference and was read as emasculation. At the same time the mantle was symbolic of the most promiscuous of Irish women, the most elusive of Irish men and of Irish resistance in general. While Irish men were imagined by English statesmen like Edmund Spencer as wild warriors who tortured and raped Irish women, these Irish women could in turn, it seemed, emasculate the Old English through intermarriage and childcare, or wander freely sowing sedition. This garment – worn by women and men alike – also undermined class and gender difference. It could hide the identity of Irish rebels and clothe unruly women. Banning the mantle could emasculate Irish men and shore up the masculinity of the colonisers, their political power and the stability of cultural difference. While Gaelic Irish and Old English culture was being suppressed in the seventeenth century, Ireland appears allegorised as a woman in poetry written in both Irish and English. In the eighteenth-century Irish poetic tradition of the *aisling*, Ireland is figured as a *spéir bhean* or sky-woman and calls the implicitly male reader to rescue her from colonial oppressors (Cairns and O'Brien Johnson 1991: 3). These different projects show the flexibility of associations between gender, sexuality and nationhood as well as prefiguring the nature of later constructions of gendered nationhood and national gender identities along lines of heroic masculinity and national motherhood.

While English versions of Ireland in the early modern period figured Ireland as a woman who evoked disgust, or desire for penetration and degradation, by the nineteenth century, English representations of Ireland as female had shifted to the model of marriage and its ideals of male affection, patronage and benevolence – but also unquestioned male discipline, authority and control. In doing so they continued to justify a colonial relationship through patriarchal gender relations. Nineteenth-century English discourses on Ireland combined Victorian domestic ideology with pseudo-scientific ethnographical analyses of racial characteristics. Matthew Arnold, following Ernest Renan's characterisation of the Irish as 'an essentially feminine race', infamously defined the Irish as Celtic and feminine (Cairns and Richards 1987; 1988: 42–51; Valente 1994: 190–1). The supposedly Irish feminine characteristics of sentimentality, ineffectuality, nervous excitability and unworldliness rendered the Irish incapable of self-government, it was argued, and thus invited strong, dispassionate and rational Anglo-Saxon rule. This marriage of races therefore simultaneously naturalised both the apparently benevolent subordination of the colonised by their colonial rulers and the subordination of women in marriage (Cullingford 1987: 1). Ireland and England and women and men were tied in a natural relationship of intimacy and inequality.

In the cultural and political claims to independence in the late nineteenth and early twentieth centuries, Irish nationalists in turn deployed gendered ideas of national character in ways which also defined the nation as female but which claimed a fierce virility for Irish men. At the same time that nationalists reworked the meaning of the colonial feminisation of Ireland, they asserted the masculinity of Irish men. In reaction to colonial racial discourses and the celebration of feminine Celtic qualities of otherworldliness in the Celtic Twilight writing of the literary revival, authors of the so-called Irish-Ireland movement asserted the masculinity of the Gael and criticised the effeminacy of both the English and misguided Irish men. This hypermasculinity inverted the colonial stereotype but retained its ideology of gender inequality. Masculinity was asserted in contrast to the femininity of women and thus demanded absolute difference between the characteristics and roles of men and women. In the literary revival of this period, writers and dramatists turned to the mythological tradition of the sovereignty goddess (Ap Hywel 1991; Innes 1993). Countless female embodiments of Ireland called on stage and page for Irish men to forsake their individual interests for the immortality of heroic self-sacrifice for the nation. In this embodiment, the female allegory was desexualised and venerated as a pure mother. Avowals of heroic masculinity were thus made alongside the celebration of dependent, passive, domestic and selfless Irish femininity. Within dominant nationalist ideology, women were elevated as producers of heroic sons, yet their quotidian, domestic, material world was dismissed as a distraction from men's senses of abstract loyalty to and endeavours for the nation. This simultaneous reverence for, and marginalisation of, women 'reinforced social and domestic sexual colonisation at the very moment it was politically overthrown' (Cairns and Richards 1987: 55). The nationalist investment in the trope of 'an idealised persona suffering historic wrongs; the sacrifice of a few in each generation to maintain this entity; [and] recurrent heroic failures to eject the invader, which culminate finally in regained independence' (Cairns and Richards 1991: 130) entailed heroic masculine sacrifice for an idealised allegorical mother figure. Conversely, women were confined to a domestic sphere that must be maintained by feminine passivity and transcended by valiant men.

Although gender and sexual norms were thus established in Ireland in the early twentieth century in the national response to the cultural and psychological effects of colonialism, they also reflected the changing class structure of rural Ireland. Post-Famine changes in patterns of landownership and practices of inheritance were closely associated with changes in attitudes to marriage and sexuality, which found support in the Catholic emphasis on sexual regulation and women's chastity and maternal role. The maintenance and improvement of small farms became a primary concern. Within this socio-economic system, known as familism, marriages were based primarily on economic considerations and on the priority of transferring the farm intact to the son chosen as most suitable. Marriages were frequently arranged, delayed or prohibited and non-marital sexuality outlawed. Illicit sexual relationships could result in unplanned marriages which would threaten the smooth transfer of property. In this period, sexuality came to be equated with matrimonial reproduction and indisputable norms of familial reproductive heterosexuality. The cult of the Virgin Mary, which flourished from the late nineteenth century – asserted in part in opposition to the Protestantism of the colonial rulers – strengthened the construction of asexual, maternal and domestic femininity upon which hypermasculinity and socio-economic and sexual regulation depended. In addition, the shift in Irish agriculture from labour-intensive mixed farming to pastoral farming resulted in greater restriction of women to strictly domestic labour and enhanced economic dependence (Bourke 1993). In efforts to secure cultural autonomy and maintain the cultural purity of Ireland after independence, women became the measure of the nation. Their idealisation as its mothers was evident in the anxieties expressed about foreign corruption of Irish women. Foreign fashions, film, literature, music and dance and foreign notions of sexual equality, it was said, undermined the home and native honour towards women and degraded Irish women. After independence, legislation restricting women's involvement in jury service and employment in the civil service weakened their status as citizens, since it curtailed women's rights to involvement in the public world of civic duty and responsibility (Gardiner 1991; Valiulis 1995). Thus the eventual ratification of women's political, social and economic subordination in the 1937 Constitution – which defined their role as maternal and femininity as essentially passive, private and domestic (Scannell 1988), implicitly enshrining heterosexuality by placing the emphasis on the reproductive family – had its origins in a complex inter-section of colonial and nationalist gendered discourses, religious belief and the economic priorities of the rural middle class.

Despite the different political motivations of colonial and nationalist feminisations of Ireland, they depend upon a shared view of femininity as weak and dependent upon male intervention. In both, female Ireland requires men to act as protectors, rulers or liberators. The condition of this female personification of Ireland – unruly or tamed and guided in colonial discourse, old and sorrowful or young, beautiful and content in national versions – depends upon her relations with men. Thus the idea of Ireland as female has been deployed in projects of subjugation and resistance. These representations of Ireland as female have been predominantly produced by men in order to stabilise versions of masculinity and naturalise their power over women. In both, the relationship between colonising or nationalist men and the gendered entity of Ireland is figured according to versions of womanhood, which define women, not through their own action, but through their sexual and familial relation to men. In these discourses Ireland is raped, seduced or married and in turn features alternatively as virgin, wanton woman, bride, mother or old woman. Nationalist versions of Ireland constructed roles and identities for Irish men through their loyal and self-sacrificing relationship towards a revered sorrowful mother or virginal figure, who would be exonerated by the political freedom of Ireland. Thus the feminisation of Ireland by English colonists and its celebration as a female entity honoured and protected by heroic Irish men are as much about the production of masculinities and femininities as they are about the making of nations. While the gendering of the nation has essentialised and simplified both gender and nationhood, women working in overlapping academic, cultural and political spheres have complicated ideas of gender, nation and identity in Ireland. The nature of their intervention in and refig-uring of the terrain of Irish identity politics are explored in the next section.

RETHINKING NATIONHOOD

Women's efforts in academic and cultural fields and their campaigns to change employment, family and reproductive and welfare law have raised crucial questions about traditional understandings of history, politics, gender and sexuality, which simultaneously stimulate new ways of thinking of nationhood. For more than a decade, the reproductive and sexual body has been the subject of vigorous and divisive moral and religious debate (Fletcher 1995; Smyth 1992a). Women's campaigns for legislative change and individual resistance to state legislation, through journeys for jobs, abortions, social and sexual freedom, have destabilised the nation by ques-tioning the kinds of gendered and sexual bodies that it proscribes. Legal disputes and personal tragedies, resulting from the legislative authority of the Catholic church in prohibiting information and access to reproductive control, raise questions about the legitimacy of the close relationship between church and state (Smyth 1995). Through individual activity and collective involvement in academic, educational, cultural and community projects, through pressure groups and formal politics, Irish women are resisting cultural as well as political marginalisation. In addition, many writers and artists have been critical of the representation of women in Irish culture and work towards alternative representations of autonomous and self-determined versions of femininity. Often they disrupt notions of essential and frequently conservative gender identities in Ireland through exploring ideas of history, geography and femininity (Cummins *et al.* 1987; Leonard 1994; *Relocating History* 1993).

Within the imaginative geography of the nation, particular places, regions or landscapes are used to construct and express senses of collective history and shared senses of belonging (see Chapter 9). Thus critical attention to the way in which the nation is conceptualised has focused on the ways in which what has come to be understood as national history has been structured through gender. In addition, both critiques of traditional nationalism and attempts to articulate different versions of cultural identity focus on the relationships between gender and geography. The examples of historical research and cultural practice which follow address both these sets of roles assigned to women in Irish history and culture and the ways in which the symbolic importance of certain places (a battlefield or the West of Ireland for example) or kinds of places (urban, rural or suburban) within the nation is bound up with notions of gender and sexuality.

Irish historians working in the field of women's history prioritise the recovery of knowledge of women's lives and experiences in the past, yet explicitly and implicitly their work addresses both the production of history in general and the particular relationship between ideas of the nation, gender and history in Ireland (Cullen 1991; Luddy 1992–3; MacCurtin and O'Dowd 1992–3). The traditional focus on a narrative of national political struggle and the political evolution of the nation-state in nationalist discourses – and within the historical establishment in Ireland – has been the subject of much criticism and debate. Yet these debates have frequently overlooked the ways in which women have been included or excluded from the narratives of the nation (Murphy 1992). Research on the actual activity of women in nationalist and feminist movements in Ireland in the late nineteenth and early twentieth centuries illustrates the resistance to and marginalisation of women's political activity and the deep tensions and fraught alliances between feminist and nationalist strategies (Murphy 1989; Ryan 1995; Ward 1983). When politically active women like Maud Gonne or Countess Markievicz have been acknowledged within conventional national histories, it is often as exceptional women whose beauty, inspirational force and relationships to men are stressed rather than their political agency. More problematically, this focus on key figures serves to mask the more widespread involvement of women in political protest. In the Land War of 1879–82, for example, women were involved in the Land League as tenant farmers, and in the Ladies' Land League as organisers and activists (Côté 1992). The customary definition of the political as collective public activity, and the neglect of social and economic history, has deflected attention away from other forms of political action and areas of experience and thus away from women's lives. By broadening the focus of research to consider non-institutional activity and redefining political action to include everyday and sporadic acts of resistance, as TeBrake (1992) has suggested, a greater sense of women's activity in this movement is possible. Large numbers of peasant women were involved in and vital to resisting eviction and maintaining boycott.

Thus historical research on women and on the gendered aspects of social, economic and political change provides a sense of the presence of women in the past in Ireland and their contribution to and experience of these processes. This means that women's lives within poor farming households of the past, for example, can be seen as spheres in which social, economic and political changes were enacted and lived on a daily level, rather than viewed as historically and politically peripheral. In recent research, Irish women in the past figure as active individuals, negotiating, resisting and supporting the dominant institutional and ideological structures of their lives in contexts of collective meaning and conflict within and between social groups, communities and classes (Cullen and Luddy 1995; Luddy and Murphy 1989). While nationalist discourses simplified and essentialised femininity, this attention to the ways in which women's opportunities and experiences have differed according to historical period, geographical location, age, class and religion, importantly undermines ideas of the uniformity of Irish women's historical experience and of the homogeneity of Irish women.

Research on women's involvement in nineteenth-century philanthropic work (Luddy 1995), paid labour, Catholic religious communities (MacCurtin 1995) and mass emigration (Rossiter 1991) provides a sense of difference within the category – women. This is heightened by attention to all those who have been further excluded from nationalist histories through their complicity in colonialism, class position, religious affiliation or on the basis of sexuality. Research on the lives of prostitutes (Luddy 1992), the wives and daughters of English soldiers and officials, Protestant and non-conformist women, and upper-class landowning women allows the power relations between women as well as between women and men to be explored, and for issues of inclusion and exclusion to remain prominent within feminist histories. It also pushes against the boundaries of nationhood. Investigating, for example, the multiple senses of identity and affiliation of women related to Irish men serving in the British army, or working as nurses in World War I, could inform contemporary identity politics in Ireland. Through insights into the diversity and mixed senses of identity of women and men in Ireland, it may be possible to rethink ideas of belonging in ways that can combine attachment to particular cultural traditions, and senses of diversity and hybridity, while maintaining a critical focus on the diverse sources of power and oppression in the past and the present.

Questions of gender, sexual and national identity emerge also in the production of popular media histories, literature, tourist promotional material, commodities and in temporary and more lasting forms of memorialisation. The destruction, appropriation and construction of buildings, monuments and statuary reflect wider debates about Irish history and revisionism but are also sites of sexual politics. Ailbhe Smyth (1991) has been critical of the ways in which monuments in contemporary Dublin – like the statue erected in 1987 of Anna Livia Plurabelle, who represents both womanhood and the city of Dublin in Joyce's *Finnegans Wake* – reinscribe the allegorical function of women in Ireland. Even the popular renaming of the statue as 'the floozie in the jacuzzi' returns to polarised, simplified and misogynist versions of femininity. This continued celebration of male-authored female allegories of Dublin contrasts with women's attempts to articulate other histories. The controversy surrounding a statue by Louise Walsh in honour of the working women of Belfast reflected the way it challenged dominant understandings of historical significance. Commemorative occasions and cultural festivals are frequently moments in which national history and national cultural heritage are both constructed and contested. In the promotional literature for the celebration of Dublin as European City of Culture in 1991, the city was presented predominantly as one of masculine literary modernism. Yet the popular definitions of historical merit or national importance do not go unchallenged (Mullin 1991). The exhibition *Ten Dublin Women* (MacCurtin 1991), for example, provided different narratives of Dublin and stories of different women's experiences of the city in ways which reveal the sexual politics of the celebration of heritage and cultural history.

These issues were apparent in an exhibition organised in Kilmainham Gaol – now a museum – during Dublin's year as European City of Culture. Entitled *In a State*, it invited artists to reflect upon the theme of national identity in a building of nationalist symbolic importance through the imprisonment of Irish insurgents from the late eighteenth century to the 1920s. In the first decades of this century, it held prisoners of the Easter Rising and Anglo-Irish war but also, during the 1922 Irish Civil War, many Republicans opposed to the treaty with England were imprisoned and some executed here. Despite the usual emphasis in the museum on national narratives of heroic male leaders, Kilmainham also housed ordinary prisoners, women nationalists and suffragists and acted as depot for convicts before transportation. Much of the art in the exhibition addressed

119

the nature of this heritage and concepts of identity in the present, hitherto masked by the single story of national history and national identity. Geraldine O'Reilly's reproductions of the prison's records, for example, register those whose lives and experiences have not been recorded in the history of the jail or more widely in the history of the state, including women imprisoned in the mid-nineteenth century for petty crime. Louise Walsh directed attention to contemporary exclusion and marginalisation in her piece, *Out-Laws, In-Laws,* in which images of snakes were superimposed on figures of gay men and lesbian women kissing. She points out that, despite the claims of equality for all citizens in the new state in the Proclamation of Independence in 1916, all civil rights are not shared by those defined as sexually deviant in contemporary Ireland. The snakes of her work echo the snakes on carvings at the entrance to the jail, which represent the serpents of crime being restrained by the chains of justice and law. They also refer to snakes as ancient symbols of creativity and sexuality and to those serpents banished from Ireland by St Patrick in Irish Christian mythology. The snakes in her photoworks are under the surface but her images of gay men and lesbian women kissing, exhibited in the site of masculine and nationalist heroics, serve to 'make them visible, numerous, struggling and unchained' (Walsh 1991: 59). Artworks like this, together with Irish gay and lesbian writing and activism, challenge the heterosexism of traditional nationalism (Boyd 1986; Marcus 1994).

These examples highlight the politics of gender within urban contexts, yet women artists and writers also deal with ideas of nature, land and rural landscapes in their explorations of embodied senses of Irish identity because of the way land and landscape have been deeply symbolic within Irish national traditions. The importance of the rural in Irish culture intersects with the gendering of nature, and through the sovereignty goddess, the land itself, as female. In Irish nationalist discourse and poetic traditions, geography and nature have frequently been combined to suggest that both Irish identity and femininity are eternally tied to the rural – to the land and earth. Suggestions that women are essentially close to nature through the reproductive functions of their bodies can be enlisted to constrain women's opportunities and define womanhood. This formula limits both what Irishness and femininity can mean. However, themes of nature, women's bodies and national cultural traditions are explored in art and writing by Irish women. Many negotiate attachment to cultural traditions and awareness of the way these traditions have been employed in constructing Irish femininity as passive, maternal, and asexual and in producing exclusive and essentialist versions of Irish identity. Their conjunction of images of women's bodies, political activism and Irish cultural traditions, which simultaneously address the meaning of femininity and national identity, can be illustrated through artwork by Alanna O'Kelly and Eavan Boland's poetic project. In different ways, both explore senses of embodied identity, history, geography and Irish cultural traditions, one through the geography of the West of Ireland, the other focusing on a Dublin suburb.

In Alanna O'Kelly's mixed-media artwork of 1992 entitled *The Country Blooms a Garden and a Grave*, images of a maternal body, the West of Ireland and Irish oral traditions are combined to express grief and criticise injustice in the past and in the present. This forms part of her ongoing project dealing with the Irish Famine of the 1840s and with the effect of loss and displacement in contemporary Ireland. In this installation O'Kelly combined imagery of the female body and the West of Ireland, a mass famine grave, and the sound of the keen – a traditional Irish ritual, a lamenting cry performed by women – in order to address the cultural memory of the Irish Famine evident in the landscape. She also sought to recover a disused and specifically Irish women's tradition in which women were active as keeners, singers and story-tellers. O'Kelly uses the keen as a political and emotional act of protest about contemporary suffering and injustice. This strategy of recovery, re-appropriation and critique also underlies her images of a maternal body. A lactating breast, filmed under water as a fine jet of milk diffuses softly like smoke, appeared on a series of video screens amongst a complex succession of images and sounds: a litany of place-names, thunder, traditional Irish song, sighs, the keen, a seagull's cry. The verdant flora of the shoreline of a mass famine grave in County Mayo contrasts with the images of death, starvation and sterility in the human bones exposed by the waves. Yet though she uses images of the female body with its connotations of natural nourishment, and rich natural marine vegetation, she does so to point to the unnaturalness of famine, caused, not by the failure of the earth, but by economic and political inequality and oppression in nineteenth-century Irish rural life. Her references to contemporary Kurdish suffering allude to the continuing political causes of famine and displacement.

Alanna O'Kelly draws on ideas of a motherly and body-centred femininity. Yet by combining the image of the lactating breast with the keen, the artwork suggests a feminist activism which posits a protesting and resisting power for past and contemporary women in Ireland, and potentially undermines the idea that women are defined through their bodies alone. The lactating breast is conventionally relegated to the private and domestic, while motherhood is deemed incompatible with political activism and artistic creativity. Yet here maternal femininity is employed as a means of protest, which disrupts the division between male and female and so challenges the understanding of motherhood and domesticity. The culture and landscape of the West of Ireland have frequently been deployed in conservative versions of Irish identity. Yet here, in contrast to the conventional associations of nature, motherhood and the rural, the rural and the feminine are combined in order to refigure women as culturally and politically active, and to criticise uneven and unequal global political and economic relations, which result in local suffering in the past and the present. This reconceptualisation of femininity is achieved through combining images of the body and traditional Irish cultural forms and cultural landscapes, which in turn changes what these traditions and places may mean.

While O'Kelly works with an already deeply symbolic geography, Eavan Boland writes of the overlooked geography of suburban life. Her subversive dialogues with the nation arose out of a sense of exclusion from the heroic and tragic stories of national resistance by men and from Irish poetic traditions, where the traditional contrast between masculine poetic self-reflective individualism and feminine maternal and collective nurture is compounded by the allegorical function of women in Irish national traditions. Rather than discard this poetic tradition because of the way women have been pacified and simplified within it, Boland writes of her own sexuality and sense of herself as a woman in order to repossess its archive of cultural energy (1996: 127). Significantly, this involves writing with a sense of embodied identity as a woman and of a place and subject-matter devalued within national poetic traditions – the suburb and its everyday, ordinary life of domestic care. Located between the literary romance of the city and the symbolic charge of the rural, the geography of the suburb and its rhythms of daily care and patterns of growth and ageing are predominantly deemed unpoetic, and politically and historically marginalised. Her focus on feminine sexuality, the experience of motherhood and on feminine ageing undermines the fixity of idealised and allegorical femininity in eternal youth, beauty and passivity. Her attention to the overlooked details of suburban domestic life grants this geography cultural and political significance. The suburb in Boland's poetry is a place in which the past and present, myth and history, continuities and differences, personal life and public politics are interwoven. Refiguring ideas of belonging through marginalised histories, geographies and embodied femininity, Boland works to 'unsay the cadences and certainties of one kind of Irishness' (1996: 94).

Through different routes, women in the arts, scholarship and politics are changing the ways in which cultural identity and difference can be understood. Although women are not defined through a return to essential acts of giving birth or childcare, the meaning of motherhood, femininity and their geographies and, in turn, history, nationhood and politics, are redefined. Thus traditional Irish cultural forms are deployed, not to reassert a single and limited notion of gender, sexuality or cultural belonging, but to change how they can be understood. The cases cited here are examples of a much wider range of artistic and social projects, which suggest that a choice does not have to be made between either progressive approaches to gender, sexuality and Irish identity or loyalty to traditional culture. Instead, critical approaches and Irish cultural traditions can mutually inform the production of alternative senses of Irishness.

CONCLUSION

Much of this chapter has focused on the specific ways in which Irish women have been positioned in relation to conventional ideas of Irishness. These accounts of the social production and cultural politics of nationhood have politicised identity in Ireland and directed attention to the way in which the definition of essential Irishness has excluded individuals and groups on the basis of gender and sexuality as well as through their cultural and religious identifications. Yet the critique of the gendering of nationhood and the curtailment of certain kinds of sexuality in Ireland has impacted upon ideas of politics and identity more widely. This challenge to the conceptions of the political and to the specific nature of Irish politics is most explicit in women's campaigns for civil rights and social justice in Northern Ireland and the Republic.

The history of Ireland and women's activism in contemporary Northern Ireland both point to the limitations of neat oppositions and single visions. Women in Northern Ireland have lived out, managed and resisted the effects of violence and social injustice in diverse ways – from maintaining families alone and visiting imprisoned family members, campaigning for housing reform and establishing community and cross-community networks to more direct involvement in sectarian politics (Kilmurrey 1987; Leonard 1991; Ward 1991; Wilford 1996). Because political action has been defined as male and military in both nationalist and unionist traditions (Benton 1995), women's efforts have been marginalised and restricted. Their concern with social issues and the violence enacted on women by men within their own communities has been ignored as secondary to the contested nature of the state (McWilliams 1995). Yet the work of women may refigure the realm of the political by forcing attention to shift away from the stalemate of sectarian politics and by changing the meaning of the public, private and political. Women's politicisation of the home, through their focus on the lived effects of violence, undermines the easy contrast between the public and private and broadens the definition of the political. The Women's Coalition of Northern Ireland, for example, is based on a cross-community network of women's groups. Its policies explicitly address all forms of inequality and deprivation and aim to shift political discussion away from rigid and irreconcilable constitutional positions, to explore new forms of democracy and political participation, and to promote dialogue and compromise. The election of Mary Robinson as President of the Irish Republic in 1990 (Smyth 1992b) and the success of the Women's Coalition of Northern Ireland in the 1996 elections to the all-party talks on the future of Northern Ireland, are both evidence of support for new kinds of politics in Ireland.

Yet women's contributions to social change and political progress in Ireland do not stem from an essential, nurturing femininity or from unproblematic alliances between women, but from their experiences of negotiating between senses of shared experiences, aims and different identities and political priorities amongst and between women (Meaney 1991; Rooney 1995; Smyth 1988). Clearly, women in Ireland vary in their attitudes and aspirations. Yet, with greater representation of women within the political arena, including political negotiations on the future of Northern Ireland, women can draw on their valuable and extensive experience of dealing with shared and conflicting political viewpoints in voluntary and community organisations. Women's experience of marginalisation, which cuts across but also differs between different classes, religious and ethnic groups, can inform and change how other problems of cultural and political differences are understood and managed.

The critical approaches discussed in this chapter do not provide simple solutions to the entanglements of gender, sexuality and the politics of identity. Yet they combine their critique with attempts to find alternative forms of creative expression that can stimulate and articulate different understandings of culture and identity. These new and evolving cultural expressions of a form of Irishness, which can mediate between senses of collective identity and diversity, and between deep attachment to cultural traditions and openness to the alternative readings of this culture, are inseparable from different claims to belonging, new forms of politics and political change. The demanding task is simultaneously to recognise the multiple forms of class, gender, sexual and state oppression in Ireland in the past and the present, consider both the specific claims and aims of women and sexual minorities in Ireland, and explore how these debates may contribute to wider reformulations of identity in Ireland. Irish feminists' focus on cultural, geographical and historical senses of embodied Irishness challenges the meaning of historical significance, politics and the imagined geographies of nation, gender and sexuality. To confuse simple, traditional, binary understandings of cultural, gender and sexual identity is to change what Irishness can mean.

ACKNOWLEDGEMENT

I am grateful to Monica McWilliams for providing extensive information about the aims of the Women's Coalition of Northern Ireland.

REFERENCES

There is an increasingly large and valuable range of literature on the issues raised in this chapter. For an extensive guide to published sources on women and Ireland see Cullinan (1995). Wide and detailed introductions to the subject are provided by Cairns and O'Brien Johnson (1991), Curtin *et al.* (1987) and Smyth (1993).

Ap Hywel, E. (1991) 'Elise and the Great Queens of Ireland: "femininity" as constructed by Sinn Féin and the Abbey Theatre, 1901–1907', in D. Cairns and T. O'Brien Johnson (eds) *Gender in Irish Writing*, Milton Keynes: Open University Press.

Benton, S. (1995) 'Women disarmed: the militarization of politics in Ireland 1913–23', *Feminist Review* 50: 148–72.

Boland, E. (1989) *A Kind of Scar: The Woman Poet in the National Tradition*, Dublin: Attic Press.

—— (1996) *Object Lessons: The Life of the Woman and the Poet in Our Time*, Manchester: Carcanet Press.

Bourke, J. (1993) *Husbandry to Housewifery: Women, Economic Change, and Housework in Ireland, 1890–1914*, Oxford: Clarendon.

Boyd, C. (1986) *Out for Ourselves: The Lives of Irish Lesbians and Gay Men*, Dublin: Dublin Lesbian and Gay Men's Collectives and Women's Community Press.

Cairns, D. and O'Brien Johnson, T. (eds) (1991) *Gender in Irish Writing*, Milton Keynes: Open University Press.

Cairns, D. and Richards, S. (1987) 'Woman in the discourse of Celticism', *Canadian Journal of Irish Studies* 13, 1: 43–60.

—— (1988) *Writing Ireland: Colonialism, Nationalism and Culture*, Manchester: Manchester University Press.

—— (1991) 'Tropes and traps: aspects of "Woman" and nationality in twentieth-century Irish drama', in D. Cairns and T. O'Brien Johnson (eds) *Gender in Irish Writing*, Milton Keynes: Open University Press.

Carroll, C. (1993) 'Representations of women in some early modern English tracts on the colonization of Ireland', *Albion* 25, 3: 379–94.

Côté, J. (1992) 'Writing women out of history: Fanny and Anna Parnell and the Irish Ladies' Land League', *Études Irlandaises* 17, 2: 123–34.

Crone, J. (1988) 'Lesbian Feminism in Ireland', *Women's Studies International Forum* 11, 4: 343–7.

Cullen, M. (1991) 'Women's History in Ireland', in K. Offen, R. Roach Pierson and J. Rendall (eds) *Writing Women's History: International Perspectives*, Bloomington: Indiana University Press.

Cullen, M. and Luddy, M. (1995) *Women, Power and Consciousness in Nineteenth-Century Ireland: Eight Biographical Studies*, Dublin: Attic Press.

Cullinan, M. (1995) 'Bibliography; Irish women', *Journal of Women's History* 6, 4 and 7, 1: 250–77.

Cullingford, E. B. (1987) ' "Thinking of her as Ireland": Yeats, Pearse and Heaney', *Textual Practice* 4, 1: 443–60.

Cummins, P., Jones B., Murphy P. and Smyth A. (1987) 'Image making, image breaking', *Circa* 32: 13–19.

Curtin, C., Jackson, P. and O'Connor, B. (eds) (1987) *Gender in Irish Society*, Galway: Officina Typographica/Galway University Press.

Finn, C. (1995) 'A question of identity', *Off Our Backs* 25, 4: 1 and 6–7.

Fletcher, R. (1995) 'Silences: Irish women and abortion', *Feminist Review* 50: 44–56.

Gardiner, F. (1991) 'Political interest and participation of Irish women 1922–1992: the unfinished revolution', *Canadian Journal of Irish Studies* 18, 1: 15–39.

Hadfield, A. and McVeagh, J. (1994) *Strangers to That Land: British Perceptions of Ireland from the Reformation to the Famine*, Gerrards Cross: Colin Smythe.

Innes, C. L. (1993) *Woman and Nation in Irish Literature and Society, 1880–1935*, New York: Harvester Wheatsheaf.

Jones, A. R. and Stallybrass, P. (1992) 'Dismantling Irena: the sexualising of Ireland in early modern England', in A. Parker, M. Russo, D. Sommer and P. Yaeger (eds) *Nationalisms and Sexualities*, London: Routledge.

Kilmurrey, A. (1987) 'Women in the community in Northern Ireland: struggling for their half of the sky', *Studies* 76: 177–84.

Leonard, M (1991) 'The politics of everyday living in Belfast', *Canadian Journal of Irish Studies* 18, 1: 83–94.

—— (1994) *Mother Ireland*, Coventry: Coventry Museums and Galleries.

Luddy, M. (1992) 'An outcast community: the "wrens" of the Curragh', *Women's History Review* 1, 3: 341–55.

—— (1992–3) 'An agenda for women's history in Ireland, part II: 1800–1900', *Irish Historical Studies* 28: 19–37.

—— (1995) *Women and Philanthropy in Nineteenth-Century Ireland*, Cambridge: Cambridge University Press.

Luddy, M. and Murphy, C. (eds) (1989) *Women Surviving: Studies in Irish Women's History in the Nineteenth and Twentieth Centuries*, Dublin: Poolbeg Press.

MacCurtin, M. (1991) 'The "ordinary heroine": women into history', in *Ten Dublin Women*, Dublin: Women's Commemoration and Celebration Committee.

—— (1995) 'Late in the field: Catholic Sisters in twentieth-century Ireland and the new religious history', *Journal of Women's History* 6, 4 and 7, 1: 49–63.

MacCurtin, M. and O'Dowd, M. (1992–3) 'An agenda for women's history in Ireland, part I: 1500–1800', *Irish Historical Studies* 28: 1–19.

McWilliams, M. (1995) 'Struggling for peace and justice: reflections on women's activism in Northern Ireland', *Journal of Women's History* 6, 4 and 7, 1: 13–39.

Marcus, D. (1994) *Alternative Loves: Irish Gay and Lesbian Stories*, Dublin: Martello.

Meaney, G. (1991) *Sex and Nation: Women in Irish Culture and Politics*, Dublin: Attic Press.

Mullin, M. (1991) 'Representations of history, Irish feminism and the politics of difference', *Feminist Studies* 17, 1: 29–50.

Murphy, C. (1989) *The Women's Suffrage and Irish Society in the Early Twentieth Century*, London: Harvester Wheatsheaf.

—— (1992) 'Women's history, feminist history or gender history', *Irish Review* 12: 21–6.

Ní Dhomhnaill, N. (1994) 'An t-Anam Mothala: the feeling soul', in J. P. Mackey (ed.) *The Cultures of Europe: The Irish Contribution*, Belfast: Institute of Irish Studies.

Relocating History: An Exhibition of Work by 7 Irish Women Artists: Vivien Burnside, Marie Foley, Frances Hegarty, Patricia Hurl, Aileen MacKeogh, Moira McIver, Alice Maher (1993), Belfast/Derry: Fenderesky Gallery at Queen's/Orchard Gallery.

Rooney, E. (1995) 'Political division, practical alliance: problems for women in conflict', *Journal of Women's History* 6, 4 and 7, 1: 40–8.

Rossiter, A. (1991) 'Bringing the margins into the centre: a review of aspects of Irish women's emigration', in S. Hutton and P. Stewart (eds) *Ireland's Histories: Aspects of State, Society and Ideology*, London: Routledge.

Ryan, L. (1995) 'Traditions and double moral standards: the Irish suffragists' critique of nationalism', *Women's History Review* 4, 1: 487–503.

Scannell, Y. (1988) 'The Constitution and the role of women', in B. Farrell (ed.) *De Valera's Constitution and Ours*, Dublin: Gill and Macmillan.

Sharkey, S. (1994) *Ireland and the Iconography of Rape: Colonisation, Constraint and Gender*, London: University of North London Press.

Smyth, A. (1988) 'The contemporary women's movement in the Republic of Ireland', *Women's Studies International Forum* 11, 4: 331–41.

—— (1991) ' "The floozie in the jacuzzi" ', *Feminist Studies* 17, 1: 7–28.

—— (ed.) (1992a) *The Abortion Papers*, Dublin: Attic Press.

—— (1992b) ' "A great day for the women of Ireland": the meaning of Mary Robinson's Presidency for Irish women', *Canadian Journal of Irish Studies* 18, 1: 61–75.

—— (ed.) (1993) *Irish Women's Studies Reader*, Dublin: Attic Press.

—— (1995) 'States of change: reflections on Ireland in several uncertain parts', *Feminist Review* 50: 24–43.

Smyth, C. (1991) 'Cherry Smyth', in R. Wall (ed.) *Leading Lives: Irish Women in Britain*, Dublin: Attic Press.

TeBrake, J. (1992) 'Irish peasant women in revolt: the Land League years', *Irish Historical Studies* 28: 63–80.

Valente, J. (1994) 'The myth of sovereignty: gender in the literature of Irish nationalism', *ELH* 61: 189–210.

Valiulis, M. G. (1995) 'Power, gender and identity in the Irish Free State', *Journal of Women's History* 6, 4 and 7, 1: 117–36.

Walsh, L. (1991) 'Thoughts on justice and snakes', in J. Graeve (ed.) *In A State: An Exhibition in Kilmainham Gaol on National Identity*, Dublin: Project Press.

Ward, M. (1983) *Unmanageable Revolutionaries: Women and Irish Nationalism*, Dingle: Brandon.

—— (1991) 'The women's movement in the north of Ireland: twenty years on', in S. Hutton and P. Stewart (eds) *Ireland's Histories: Aspects of State, Society and Ideology*, London: Routledge.

Wilford, R. (1996) 'Women and politics in Northern Ireland', *Parliamentary Affairs* 49, 1: 41–54.

7

IN SEARCH OF ETHNICITY IN IRELAND

Michael A. Poole

INTRODUCTION

This chapter had its genesis in a throwaway remark at a Belfast conference, at which a group of English experts sought to enlighten a Northern
Ireland audience about the opportunities for analysis from the 1991
census of population. The comment was simply that, unlike Great
Britain, Northern Ireland had no ethnic question on its census form.
While the statement was superficially absurd to anyone familiar with the
social science literature on Northern Ireland, more charitably it implied
that the specific ethnic question asked in Great Britain is not used in the
Northern Ireland census – nor, more fundamentally, are the speaker's
perceptions of ethnic categories.

The paradox has its roots in the use of language. Here is a term which is
understood to mean one thing by a social scientist from Great Britain, but
something quite different by an analyst of the Northern Ireland situation.
The kind of identities conventionally classified as ethnic in Britain are, to
quote some of the pre-coded categories from the census form, 'White',
'Black-Caribbean', 'Indian', and 'Pakistani'. These labels illustrate a strong
concern with geographical family origin but, above all, an obsession with
the visual cues of 'race' which seem to dominate British interpretations of
ethnicity. For example, Smith (1989: 13–14) deliberately bypassed the
ethnicity model in her analysis because, she points out, 'in the British
context . . . ethnicity has become a euphemism for race.'

The kind of ethnic minority groups identified in the British census
categories are conspicuously rare all over Ireland. For example, McVeigh
(1992: 32–3) estimates that the island has a total of 20,000 people of
Chinese, South Asian, African or Caribbean origin: thus the population of
Ireland is 99.6 per cent 'white'. However, when the term 'ethnic' is
applied in Ireland – especially in the context of the North – reference is
almost invariably being made to the distinction between Catholic and
Protestant. True, the same division may also be described by a whole host

of other labels, such as political, national, religious, denominational, sectarian, tribal and even racial (Benson and Sites 1992; Macourt 1995; McKernan 1982). There are hybrid labels, too, which reflect the multidimensionality of the division – ethnonational, ethnopolitical, ethnoreligious and politico-religious (Boal and Livingstone 1984; Cairns and Mercer 1984; Coulter 1994; Cecil 1993). As the predominant prefix of these hybrids suggests, however, ethnicity is, to a large extent, their common denominator, reflecting the increased momentum over the last quarter-century, at least in the specialist academic literature on Northern Ireland, towards downplaying the other epithets and interpreting the fundamental Catholic–Protestant divide in ethnic terms.

Even then, however, these two interpretations of ethnicity – specified by 'race' and religion respectively – do not exhaust the full range of possible social markers that define this rather flexible term. There is a growing, though not uncontested, opinion that Ireland's Travellers – tinkers or itinerants in the parlance of earlier generations – constitute a distinct ethnic group (Helleiner 1995; Calmy 1987). However, since they number only about 22,000 in total (MacLaughlin 1996: 43), they will be omitted from the empirical analysis in this short chapter. So, too, will the racially defined groups of non-European origin, who are even less numerous. Rather, I will focus on the potential of religion in defining ethnicity.

The word 'potential' has been chosen deliberately. While the ethnic interpretation of religious division has become relatively conventional in academic writing on Northern Ireland, it is much less normal to apply this rendition in the Republic. One rare exception is provided by Bowen (1983), who employs the construct, albeit only occasionally, in labelling the Protestant minority. This contrasting usage on either side of the border indicates a geographical variability in the interrelationship between religion and ethnicity, emphasising again the flexibility of the way in which the meaning of the latter is constructed. Moreover, the composition of ethnicity is subject to fluidity over time, as demonstrated by the nineteenth-century convergence of Scots Presbyterians and English Episcopalians into a shared Ulster Protestant and unionist identity (Pringle 1985). Therefore, ethnic categories must not be conceived as fossilised entities extending either backwards or forwards chronologically. It is also notoriously difficult to disentangle this 'complex and dynamic process' (Nanton 1992: 284) from certain allied concepts, most notably, nationalism. The latter, a much more commonly used term in Ireland, is also related to religious affiliation. The result is that ethnicity, nationalism and religion define the three corners of a triangle, and this chapter seeks to explore the totality of the ensuing complex of linkages (Figure 7. 1). The underlying motive is to assess the significance of ethnicity in the creation of social identity in Ireland and, moreover, to ask whether – if it is present at all as a dimension of social differentiation – its importance varies from one part of the island to another,

129

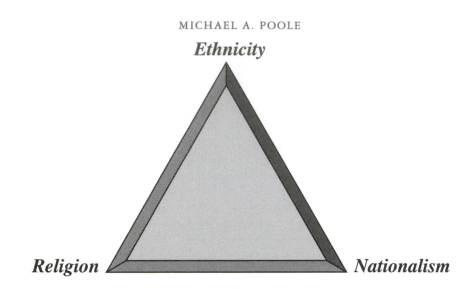

Figure 7. 1 The triangle of ethnicity, nationalism and religion

and through time. The spatial arena analysed is therefore the whole island of Ireland, particular emphasis being placed on the contrasts between North and South.

THE JUSTIFICATION AND CONCEPTUAL BASIS OF THE ETHNIC MODEL

Before embarking, however, on the analysis that attempts to answer these questions, the questions themselves require some justification. After all, does it matter whether it is valid to apply the model of ethnic identity to Ireland, and is it of any significance to know the specific guise in which ethnicity manifests itself on this island? It is possible to argue that the investigation of the applicability of the model merely reflects a chronic academic urge to classify. However, ethnicity involves not just a codifying label but an entire package of both theoretical and empirical research. It is the empirical context that will be emphasised in this chapter, one that has witnessed in excess of 3,000 dead in twenty-five years of political violence in Northern Ireland (Sutton 1994). Moreover, these killings are only the most vicious manifestation of a far wider diversity of personal catastrophes – severed limbs, mental trauma, refugee flight, destruction of livelihood – which are symptomatic of the gravity of the underlying socio-political problems. Thus it is vital to understand whether the Northern Ireland conflict is, for instance, a religious division between the theological categories of Catholic and Protestant, or a political one between two nationality groups whose religious designation is no more than a convenient label.

The former view is inherent in what McAllister (1982: 330) calls the popular belief that Northern Ireland is a historical anachronism, 'sustaining a sectarian conflict over issues which were largely resolved in the rest of western Europe centuries before'. The significance of such a purely theological interpretation of the Northern Ireland situation is that the search for analogies to aid comparative analysis tends to focus on the relations between religious groups elsewhere in the world. This leads people to observe, for example, that Catholics and Protestants are no longer in serious conflict in Great Britain, the Netherlands or the United States and then to ask why the population of Northern Ireland cannot live in similar harmony. Such reasoning is not only simplistic but also misunderstands the conflict by seeking its explanation in a supposed cultural abnormality of the Province, rather than in the tragic imperatives of its social environment (Ruane and Todd 1991). Indeed, it is an approach that can easily lead to a racist analysis, resorting to some stereotypical notion of Irish aggression.

On the other hand, alternative models of the Northern Ireland conflict lead to the choice of totally different geographical analogies. These alternatives tend to focus on the other two corners of the ethnicity–nationalism–religion triangle. For example, an ethnic depiction of the Northern Ireland conflict suggests a comparison with race-based ethnic divisions in Great Britain and the United States, or with linguistic cleavages in Belgium and Spain. If interpreted as a national conflict, implying the rejection of sovereign state boundaries, comparisons are prompted with Israel and Palestine, or Sri Lanka, or with the fracturing of the former Yugoslavia and Soviet Union. The contrasting nature of these possible parallels – and the associated differences in the degree of gravity of the conflicts involved – serve to emphasise that the way we choose to conceptualise the Northern Ireland conflict is of fundamental importance. It has a profound impact on the ways in which we think about its mutually antagonistic groups, our evaluation of the problem's apparent intractability, and the international contexts in which we set our thinking on this Irish conflict. Nor are these questions restricted to the contemporary conflict in Northern Ireland alone. All of Ireland has a long tradition of politically motivated violence, raising the possibility that, if the recent troubles in the North can be labelled as 'ethnic', then earlier all-Ireland violence might also be categorised in the same way. This, in turn, leads to the issue of the resolution of those ethnic divisions in the post-partition period south of the border.

Although I have sought to emphasise the complexity of ethnicity, especially its flexibility with respect to social markers and its fluidity over both time and space, no definition has yet been provided. As Sillitoe and White (1992: 143) observe, 'when we speak of an *ethnic group*, we mean a socially distinct community of people who share a common history and culture and often language and religion as well.' Three basic strands can be isolated from this statement: the activity segregation which gives rise to the socially distinct community; the myth or actuality of a common perceived historical and cultural origin distinguishing the group from others; the delimitation of the group by key social or cultural markers such as language and religion. A definition of ethnicity framed in these terms implies a model of cultural separateness, which must be differentiated from the related concept of nationalism. The distinction has been discussed in the Irish context by Gallagher (1995), who quotes Smith (1991: 8–15) as distinguishing 'between a Western "civic" model of the nation, entailing "a community of people obeying the same laws and institutions within a given territory" and an Eastern "ethnic" model, emphasising a community of birth and native culture.' This is a theoretical distinction which will play a central role in my argument, its importance underlined by Kellas's observation (1991: 51) that ethnic nationalism is exclusive because it shuts out those who are not part of the ethnic nation. Consequently, social exclusion is a very significant consequence of one form of nationalism but not the other.

Being excluded from, or peripheral to, the mainstream of society need not be a permanent state, however, and the American literature on ethnicity has always been especially concerned with this type of change. In particular, it has focused on the degree to which European ethnic groups have been subject to a process of assimilation into mainstream white American society. This implies that distinctive ethnic identities declined over successive gener-

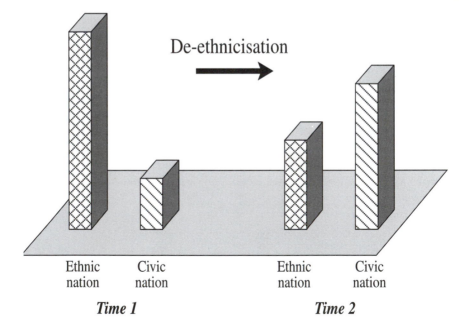

Figure 7. 2 De-ethnicisation and the shift from ethnic to civic nationalism

ations, leading to an expansion of the population who identified primarily with the civic conceptualisation of the American nation. However, the North American analogy also introduces two complications concerning the conceptualisation of ethnicity and its application to the two parts of Ireland. First, the simple idea of assimilation implies that there may be shifts over time between ethnic identification and national identification, and this idea can be linked to Smith's dichotomy between ethnic and civic nationalism to generate the hypothesis that there may be identity shift between these two forms of nationalism. If this is a movement from the ethnic form to the civic, then it may indicate that many people lose ethnicity over time or that families lose it from one generation to the next (Figure 7. 2). In this case, assimilation may be said to take the specific form of *de-ethnicisation*, which is perhaps analogous to the process of secularisation in the religious context.

The second complication arising from the introduction of the American material is the recognition that ethnic identity is not an attribute which is simply present or absent, for people may have it to varying degrees. This means that ethnic attachment can be hypothesised to change over time, and indeed there may be both short-term and long-term components to this. Ethnic identity is a feeling subject to ebb and flow: events and circumstances can strengthen or weaken it, so people may appear de-ethnicised one day but seem to have returned to their ethnic roots the next. However, there may be more consistent trends over longer time-periods – both in one person and between generations. The variable strengths of ethnicity have important implications for de-ethnicisation, for it means that it may be mistaken to expect a simple de-ethnicised bloc in society. Instead, the process may mean that there are certain groups in society who are rather more de-ethnicised than others, with degrees of difference involved rather than clear separation. The whole argument here about de-ethnicisation and the variable strength of ethnicity is consistent with Nanton's (1992) warning of the dangers of rigidly simplistic and unchanging ethnic classification in the context of, for example, census analysis. Having now demonstrated both the relationship between ethnicity and nationalism, and the strength of ethnic identity itself, observing, in each case, the fluidity and complexity of these concepts over time, it remains to apply these ideas to the Irish context.

ETHNIC IDENTITY IN NORTHERN IRELAND

It is not just the language of Great Britain which has a disinclination to refer to the Northern Ireland conflict as ethnic. Local political discourse in the Province itself tends either to resort to the evasive euphemism, 'community' – as in 'community conflict' and 'community relations' – or to label the conflict as religious or sectarian. Conversely, the expression 'ethnic' is often employed locally in the British way – not surprisingly in view of the influence of the London-based media. Thus a recent University of Ulster survey found that, when asked to state their ethnic identity, three-quarters of students gave an answer which included the word 'white', while only one-quarter included a religious label as even part of their response.

However, whereas academic and popular language in Britain largely co-incide in using ethnicity as a polite euphemism for 'race', there is a lexical divergence in Northern Ireland between popular vocabulary and that of academic specialists. Much of the responsibility for this can be credited to F. W. Boal who, since the start of the Troubles, has emphasised that the Catholic–Protestant split represents so fundamental a schism, with implications for social interaction far beyond religious activity alone, that it must be interpreted in ethnic terms (Boal et al. 1976). Most contemporary analysts of the Northern Ireland conflict now accept this reasoning. Thus Brewer (1992: 356) describes ethnicity as now 'the most popular portrayal' of the Catholic–Protestant dichotomy, 'with the groups being seen as ethnic ones socially marked by religion', a conceptualisation which he clearly supports. Similarly, Wallis et al. (1986: 3) state that: 'There can thus be no real doubt as to Catholic and Protestant communities constituting ethnic groups.' This view reflects the conclusion that the obvious religious difference is supplemented by both fundamental political contrasts and substantially segregated activity systems, especially in education and kinship networks, as well as by a clear consciousness of distinct and 'mutually antipathetic' histories. Clearly evident in this argument is each of the three key strands present in the definition of ethnicity cited above.

Only rarely do writers on ethnicity in Great Britain acknowledge this Northern Ireland application of the concept, and even less frequently do they integrate the race-based and religious interpretations to derive an overview of the UK's ethnic relations. This does much to explain an important point made in the social policy context by Osborne (1996: 197). He uses the expression, 'intellectual disengagement', to refer to the tendency exhibited by British social policy analysts in general – and writers on equal opportunities in particular – to shun the experience of policy debate and implementation in Northern Ireland. Labelling the Province's conflict as 'ethnic' should stimulate the making of intellectual connections between the principal social division in Ulster and the differentiation between what are called 'ethnic groups' in Great Britain. However, it seems that the focus on race is so obsessively single-minded there that the Northern Ireland division is automatically regarded as being totally unrelated. Osborne emphasises that, ironically, researchers further afield appear to have no difficulty in making the necessary intellectual connection, the Province regularly being depicted as a valuable case-study of the more general phenomenon of ethnic conflict (see, for example, Engman 1992; McGarry and O'Leary 1993; Peach et al. 1981; Samarasinghe and Coughlan 1991; See 1986).

Evidence for de-ethnicisation

Notwithstanding the merits of meaningful comparison which the ethnic interpretation endows upon analyses of Northern Ireland society, it does, at least at its most elementary level, imply a very simple perspective. In conceiving Northern Ireland as a dual society, divided between Catholic and Protestant, there is a danger that these are viewed as two undifferentiated monoliths (Boal *et al*. 1991). However, simplicity is not necessarily synonymous with being simplistic and, in certain respects, such a dual view is largely valid. For example, integrated schools educate only 2 per cent of the Province's children, with the remainder attending essentially segregated schools (see pp. 163–4). Similarly, intermarriage between Catholics and Protestants accounts for just 6 per cent of all married couples, and between only 4 and 10 per cent of voters give first preference to the self-consciously 'middle of the road' Alliance Party (Compton and Coward 1989: 186; Mitchell 1995: 780).

Thus there is only limited evidence for the existence of anything more than a small 'third force' in Northern Ireland society – a 'de-ethnicised' group, in the language introduced earlier. Moreover, this group exists separately with respect to each criterion employed in defining ethnicity. Consequently, the de-ethnicised element in terms of one marker need not be the same people who are defined with respect to another. For example, Alliance supporters are certainly not overwhelmingly the partners in mixed marriages. There is thus no single de-ethnicised bloc but, instead, a much more complex pattern comprising a whole series of separate 'third-force' groups, displaying relatively little intercorrelation with each other. This is clearly demonstrated by observing the distinct geographical distribution of de-ethnicised groups. For example, the concentration of Alliance voters inland from both shores of Belfast Lough (Douglas 1989: 78–81) may be contrasted with the disproportionate location of mixed marriages in a broad north coast belt around Coleraine. Both these patterns, in turn, differ from the spatially rather uniform distribution of integrated schools (Stephen 1993: 23).

There is a direct analogy between this third-force concept in the ethnic sphere and the secular population uninvolved with church attendance. Indeed, since religious activity is one of the criteria which define ethnicity, secularisation is itself a characteristic of the de-ethnicised population. But, once again, we have a third force which is only partly correlated with other de-ethnicised groups. In fact, some of the most violently passionate advocates of a specific ethnic identity are not church attenders at all. Thus the old Ulster joke-question about whether a non-believer is a Catholic or Protestant atheist is not just an inquiry about the individual's childhood or family background. It is a double-barrelled question simultaneously aimed at ethnicity and religion, for it is perfectly possible to be theologically atheist or agnostic but ethnically Catholic or Protestant. Secularisation is therefore not the same as de-ethnicisation, a point borne out, moreover, by the geographical evidence, which reveals a further pattern different from the distribution of integrated schools, Alliance voting or mixed marriage: for both Catholics and Protestants, church attendance is lower in Belfast than elsewhere in the Province (Moxon-Browne 1980: 22).

The cumulative effect of this evidence on the disparate nature of third-force signifiers supports the suggestion that there is no simple, de-ethnicised bloc in Northern Ireland society. Instead, many people are de-ethnicised in relation to at least one of the many components that define ethnicity – but, in many cases, no more than one. These observations on behaviour show an instructive correspondence with the attitude-oriented contention of Ruane and Todd (1992: 93) that 'individuals who hold moderate views on one issue may hold far from moderate views on another'. Both types of evidence support their argument that, because Catholic and Protestant identities each have a multicentred structure, virtually everyone is closely bonded to their own group on at least one of the multiple dimensions involved.

Ethnicity and nationalism

Having established that there is value in applying the ethnic model to Northern Ireland society – as long as complexities like de-ethnicisation and the variable strength of identity are recognised – I now turn to the relationship between ethnicity and nationalism. The traditional claim of Irish nationalism to embrace the population of the whole island is obviously irreconcilable with the vociferous and often violent determination of Northern Ireland Protestants to resist absorption into a state ruled from Dublin (Whyte 1990: 191). On the other hand, it may be valid to view the Catholic minority of the Province as part of an Irish nation, especially one defined by the Catholic, Gaelic nationalist myth, which was so vigorously promoted in the Irish Free State after independence. Gallagher (1995: 718) regards the cross-border unity of this Irish Catholic nation as uncontentious, but considers it necessary to scrutinise closely 'the question of how to describe those who define themselves out of the Irish nation'. After evaluating six alternative models of national composition, he states a preference for the 'three nations' perspective, arguing that the Irish nation is accompanied by both an Ulster Protestant nation and part of the British nation. The latter is, in terms of Smith's typology, a western civic nation based on allegiance to institutions, whereas the Ulster Protestant entity, like Irish Catholic nationhood, has much more in common with the ethnonationalism of Eastern Europe.

If Gallagher is correct about the 'three nations' model (see pp. 201–2 for an alternative viewpoint), this leaves Northern Ireland with two ethnonational groups – Irish Catholics and Ulster Protestants – together with a third group which is, in one sense, non-ethnic. Certainly, this last category has no clear ethnic identity deriving specifically from its nationality, because 'British' is a civic supernationalism embracing a number of separate ethnonationalisms (Kellas 1991: 52). On the other hand, there is a fluidity of identity – both between ethnic and non-ethnic Protestants, and between ethnic and non-ethnic Catholics – which responds to changing political circumstances. People react, for example, to the latest perceived outrage – by paramilitaries or state security forces – or to an exercise in social bonding like a well-orchestrated hunger strike or a particularly provocative sectarian parade. Therefore, at times of crisis, the full set of three nations are almost completely squeezed into two ethnic nations, leaving only a tiny residue with a basic British identity.

This evidence for the variable strength of ethnic identity is important because it implies that the model of Northern Ireland society as two ethnic groups is not, in fact, invalidated by the implications of Gallagher's 'three nations' model. Indeed, this conclusion follows, not only from the crisis-driven retreats to a basic ethnonationalism but from the other dimensions of ethnicity – like education, marriage and religious practice – which are less strongly or only weakly related to national identity. Thus even the individual who is normally non-ethnic in terms of nationhood still has an ethnic identity derived from at least some of these other dimensions, since de-ethnicisation coexists side-by-side in most people with one of the forms of ethnicity. In fact, because the group with a civic British national identity is primarily – but not exclusively – Protestant in religious terms, the majority of its members have a Protestant ethnic identity, leaving a minority as part

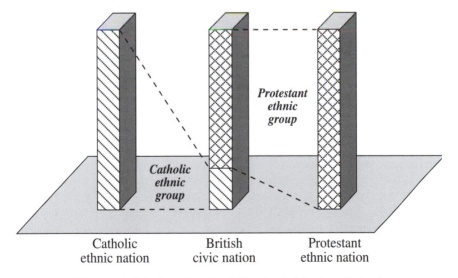

Catholic   British   Protestant
ethnic nation   civic nation   ethnic nation

Figure 7. 3 Ethnic and national identity in Northern Ireland

of the Catholic ethnic group (Figure 7. 3). In ethnic terms, Northern Ireland is therefore a dual society, despite the existence both of a non-ethnic nation and of all the other forms of de-ethnicisation.

ETHNIC IDENTITY IN THE REPUBLIC OF IRELAND

The flexible approach to modelling ethnicity also helps us to understand the incidence of ethnic identity in the Republic of Ireland. It provides a warning against trying to assess, for example, whether Protestants are, or are not, an ethnic minority in the state. Instead, it focuses attention on measuring the extent to which they form an ethnic minority at any one time. Irrespective of their ethnicity, there is no doubt that Protestants constitute a minority. Of those people who responded to the religious question in the 1991 census, 4 per cent were Protestant and 94 per cent were Catholic, while almost all the remainder claimed to have no denomination (Central Statistics Office 1994: 10). This immense Catholic majority gives the Republic an exceptionally high degree of religious homogeneity. Indeed, in their notably comprehensive survey of the entire European national and ethnic scene, Krejci and Velimsky (1981) observe that the Republic of Ireland is one of the very least divided of the continent's states in terms of either ethnicity or religious affiliation among the native population: in fact, Luxembourg and Portugal are listed as the only comparable EU countries. This homogeneity is translated into the political domain, the Irish Republic being one of the very few states within the Union which has neither a political party based on an ethnoregional support base, nor a party of the extreme right stimulated by anti-immigrant racism (Lane and Ersson 1987: 94–105).

The extent to which the Republic of Ireland is a homogeneously Catholic state has increased significantly since partition, for the new country was 90 per cent Catholic and 10 per cent Protestant at its last pre-independence census in 1911 (*Census of Ireland 1911*: 296–7). This was a much larger majority group than the corresponding 66 per cent Protestant population in Northern Ireland, but both new political units chose to confuse democracy with the concept of majority rule, thus ending up – each in their different ways – with uninterrupted majority dictatorship. This is a label which Lijphart (1975: 94) has applied to Northern Ireland, with its perpetuation of a single, all-Protestant party in regional government from 1921 until the advent of Direct Rule in 1972. However, it can also be applied to the state ruled from Dublin, where the large Catholic majority was viewed as legitimating the uniform imposition of a Catholic ethos. This was done by manipulating the legal system to shore up 'monopoly Catholicism', to use Fulton's evocative phrase (1991: 133). As Pringle (1989: 42) has emphasised, 'Catholic Church doctrine became enshrined within civil legislation on issues such as censorship, divorce, contraception, and abortion [and thereby] the Catholic majority tended to alienate non-Catholics from the mainstream of southern society.'

The result was a suppression of civil liberties, albeit affecting the state's entire population and not targeted specifically at the Protestant community – unlike the political exclusion of the Catholic minority in the North by a majority determined not to share power unless it had absolutely no alternative. However, the effect of the Catholic ethos has been to heighten the Protestant community's sense of peripherality to the core of the Republic's society and its nationalist identity of a Catholic, Gaelic Ireland. This exclusion of Protestants dates back to the creation of the nationalist construct in the later nineteenth century (Girvin 1986: 4–5), although it was particularly accentuated in the months before and after independence in the early 1920s when local campaigns of violence and persecution were waged against Protestants and their property. For example, Hart (1993, 1996) has documented the high degree to which Protestants were over-represented among the victims of IRA murders in County Cork in 1920–3, but he emphasises that this was in company with former members of the security forces as well as Travellers, the mentally disabled and sexual deviants. What all these groups had in common was that they were outsiders – minorities excluded from the fairly respectable ranks of an Irish nationalism which was distinctly bourgeois as well as ethnic (MacLaughlin 1996).

This social exclusion means that Protestants in the Republic of Ireland formed both a religious and an ethnic minority. As Kennedy (1988: 152–3) insists, Protestants have to be perceived not just as a religious group but also as a cultural and political minority. Although accepting that Protestants were assured of religious toleration, he claims that they were treated less liberally as a cultural minority – for example, in relation to the compulsory Irish language that was imposed in schools as part of the nationalist package (Akenson 1975). As a political minority, Protestants were weakened, too, by being eased out of their original deliberate over-representation in the Senate and through their falling numbers in Dáil Éireann as the mechanics of the electoral system were altered to their disadvantage (Bowen 1983).

The dominant nationalist ideology of the state therefore defined a monolithic Catholic core, leaving a marginalised periphery of non-Catholic communities – secular as well as Protestant and non-Christian. In the context of Smith's distinction between civic and ethnic nationalism, this implies that Irish nationalism is essentially ethnic. The tendency for this type of nationalism to embrace exclusivity by shutting out minorities means that Protestants have suffered in this respect, even though a relatively high average prosperity – resulting from their occupational profile and landownership – distances them from the conventional, deprivation-oriented concept of a socially excluded group.

Civic nationalism and the Protestant minority

It could be argued, of course, that southern Protestants shut themselves out from the new Irish state because their ethnic and national allegiance lay elsewhere. Indeed, all the sobriquets commonly applied to them – the 'garrison', the 'Ascendancy', 'Anglo-Irish' and 'West Britons' – carry some implication of ethnicity in the form of a link with the larger island to the east. However, such a connection was devastatingly weakened by the bitter sense of betrayal which 'snapped the emotional bond' with Great Britain (Buckland 1972: 282). In addition, the Protestant community felt itself powerless to do anything but accept the Irish Free State as a *fait accompli*. On the other hand, southern Protestants could hardly have been expected to be overnight converts from British to Irish nationalism. Thus there is a poignant significance in the expression 'ex-unionist' which soon gained currency south of the border (Akenson 1975: 110), pointedly implying as it did the ambiguity of Protestant national identity.

This status facilitated what Lyons (1967: 99–100) has famously described as a retreat into a 'ghetto mentality', involving only limited contacts between the Protestant community and its Catholic co-citizens. An insularity was maintained for many decades not only in social life – especially, as far as possible, in marriage ties – but also in economic matters such as employment practices (Bowen 1983). Such voluntary segregation is one of the classic hallmarks of ethnic identity, and it reinforced the effects of invol-

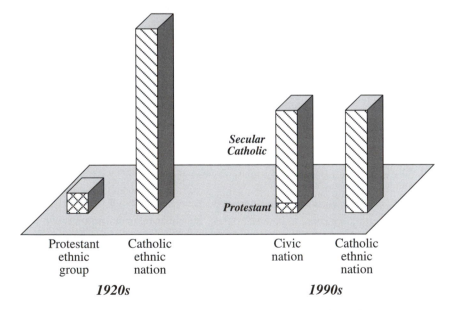

Figure 7. 4 Ethnic and national identity in the Republic of Ireland:
1920s and 1990s

140

untary social exclusion to make southern Protestants – despite their internal differences – a socially distinct community sharing a common history and culture. In other words, they constituted an ethnic group, identifiable by their religion. However, since they did not behave as a community with a distinctive national aspiration, they did not, unlike their Catholic fellow countrymen, constitute an ethnonational group (Figure 7. 4).

In later decades as a new generation – one which had not been socialised into British identity before independence – grew up, a drift began towards accepting Irishness as a national identity rather than a regional form of Britishness. This probably explains why Bowen found that his Protestant respondents, interviewed in 1973–4, differed in their self-perception as 'Irish' or 'British' according to whether they were over or under sixty years old. Only those born after c. 1914 were politically socialised after independence in 1922 and, in considering the growth of Irish identity at the expense of feelings of Britishness, it is easy to forget that it was not until about 1965 that a majority of Protestants of voting age had been born after 1914 (Central Statistics Office 1977: 28).

This Irish national identity, which did eventually supplant the post-British southern Protestant status of a people in limbo, was a civic nationalism – not ethnic – for it was tied to the *de facto* boundaries of the new state much more than was Irish Catholic nationalism. After all, the latter traditionally identified with the entire island, embracing the Catholics of Northern Ireland, whereas most southern ex-unionists never had much empathy with the Ulster population, Catholic or Protestant (Davis and Sinnott 1979: 109–10). As southern Protestants have embraced civic nationalism, the insularity of their social life has declined, partly in response to declining numbers, but also as a consequence of the progress of secularisation in both religious communities. In the economic sphere, too, there is a much lower intensity of workplace segregation. The result is that the Protestant population is now substantially less of a distinct ethnic group, thereby further consolidating its newly enhanced sense of Irish identity.

Civic nationalism and the Catholic majority

Just as importantly, however, it may be argued that there has been a significant shift in the national orientation of the Catholic majority, for, in the words of Garvin (1988: 103), 'the nineteenth-century synthesis of nationalism and Catholicism is very gradually coming apart at the seams'. Consequently that majority is now moving towards the civic nationalism, which was arguably pioneered in the southern state by the Protestant community. After three generations functioning as a separate political system, together with an acute awareness of twenty-five years of violence in Northern Ireland, the commitment to an all-Ireland identity has waned among the twenty-six counties Catholics (Douglas 1989: 62). This means that the sense of loyalty accorded specifically to the *de facto* political entity – with the implied acceptance of partition – has correspondingly waxed. For example, survey research, conducted as long ago as 1983, showed that a substantial minority of respondents – 34 per cent – defined the Irish nation as being limited to twenty-six counties (Cox 1985b: 36–7).

None of this prevents the flourishing survival of a hustings rhetoric favouring Irish unification, but then southern politics has always needed the issue of partition as a grievance, much more than it ever wanted an end to the publicly denigrated *status quo*. There is a strong rationale to this admittedly duplicitous attitude, for 'a united Ireland, especially in a unitary state, would entail the destruction not only of Northern Ireland . . . but also of the Irish Republic as it has become. Uniting Ireland means an end not to *one* Irish regime but to *both*' (Cox 1985a: 442). Above all, the southern state's claims to some degree of political and religious homogeneity and harmony – the source of much of its legitimacy – would be utterly destroyed by having to include the whole of Ulster.

In addition, the progress of secularisation in the Republic of Ireland has weakened the commitment to Catholicism, including its role as a component of an ethnic national identity. The shift away from traditional church teaching was certainly shown by the large minorities voting for change in the abortion and divorce referenda of the 1980s, a trend culminating in the approval in 1995 of the right to divorce (Adshead 1996). The exceptional narrowness of the majority, however, shows how evenly divided the electorate is on secularisation, and the bitterness of the passions aroused demonstrates how deep is that division. Furthermore, the trend towards both tolerance of a genuinely plural society and the secularisation that is eroding the Catholic hegemony does not mean the disappearance of Catholicism as either a religious or a political force. Indeed, the European evidence suggests that, in countries like France, Italy and Spain, conflicting Catholic and secular camps coexist for decades or even centuries – each entrenched, moreover, in their own geographical regions.

Barriers have not been removed, therefore, either between Catholic and Protestant or between Catholic and secular, but the wider significance of these hurdles, in terms of ethnicity and nationalism and hence in terms of politics, has certainly declined. It is true that secularisation must not be confused with de-ethnicisation, for it is just as possible to be an ethnically committed, but religiously lapsed, Catholic in the South as it is to be a Protestant atheist in the North. However, the fact that the tendency to define the Irish nation in twenty-six county terms is greatest among young voters and in the larger towns is instructive, for these are the very groups with the highest propensity to weak religious commitment (Cox 1985b: 37–8). Not only is this evidence for the correlation of civic nationalism with secularisation, but, unless the young turn more to religion as they get older, the age factor will combine with expanding urbanisation to generate a relentless momentum towards both civic nationalism and secularisation. The movement of opinion over even a single decade is indicated by the rise in the support for divorce from 36.5 per cent in 1986 to 50.3 per cent in 1995. This shift, in turn, has major implications for the Protestant minority, for there is increasingly an alliance in the Republic's society between Protestants and the more secular Catholics, who are thereby placed in the opposite camp from the traditional Catholic ethnonationalists (Considère-Charon 1995: 188). It is this division, and the realignment that it implies (Figure 7. 4), which has almost ended the social exclusion of the Protestant community and thereby done much to eliminate the distinctive ethnic identity which was, to a large extent, thrust upon it.

CONCLUSION

I have argued that, because of the adoption of the British tendency to use the expression 'ethnicity' as a euphemism for 'race', most Irish people are reluctant to describe the social division of either part of their island in terms of the language of ethnicity. Nevertheless, it may be argued that both political units have experienced profound ethnic splits, based on the link between religion and nationalism. Ethnic cleavage has been most apparent in Northern Ireland, as the comparative international literature has recognised for a quarter of a century. The obvious twofold division, identified by Catholic and Protestant marker-labels, has been increasingly complicated by de-ethnicisation, but this has failed to create a distinct non-ethnic middle bloc in Ulster society. Instead, there are separate dimensions of de-ethnicisation, and individual people exhibit a simultaneous mixture of ethnic and de-ethnicised characteristics. Consequently the strength of ethnic identity differs from person to person, but virtually everyone remains at least partially ethnic by virtue of identifying more strongly with one side than the other. Therefore, in spite of the argument for three nations in Northern Ireland, there remain only two ethnic groups, implying the existence of two ethnic nations in the Province, each separated by a notoriously blurred boundary from a British-oriented civic nation.

The situation is more clear-cut in the Republic, with only a single ethnic nationalism – one to which an overwhelming majority of the population could, at one time, owe allegiance by virtue of its Catholicism. However, the Protestant minority formed an ethnic group, too, albeit not an ethnonational one. Protestantism, in fact, acted as the vanguard for the development of a civic nationalism in the twenty-six county state. Now, however, the progress of secularisation among the Catholic community has added much greater strength to this form of nationalism, a trend that leaves the southern state's nationalism divided into two camps – ethnic and civic – though the boundary between the two is as 'fuzzy' as in the North. These changes have not altered the fundamental structure of ethnicity in the Republic of Ireland, for the two basic religious groups can still be viewed as ethnic entities. But the strength of ethnic identity has changed, with de-ethnicisation having advanced much further among the supporters of civic nationalism – Protestant and secularised Catholic – than those clinging to the more exclusionary Catholic ethnonationalism. This division is eliminating the traditional religious and ethnonational near-homogeneity of the Irish Republic, so, ironically, de-ethnicisation – because it is only partial – is generating increased ethnic heterogeneity.

The result is that both parts of Ireland are divided internally along an axis involving ethnicity, so this form of cleavage is fundamental to the creation of social identity throughout the island. However, despite a common de-ethnicisation trend and notwithstanding some recent involvement for the Dublin government in Northern Ireland's internal affairs (Morrow 1995), the specific nature of the nexus of ethnicity, nationalism and religion in the two political units is so different that they are drifting even further apart. The Irish Republic is, in one sense, heading in the direction of France, Spain and Italy, with an almost uniform nominal religious allegiance masking a split between the secular and the religious – although the similarity should not be exaggerated, since there is more in common along this purely theological dimension than along the ethnic. However, in more specifically ethnic terms, there is again a parallel with trends in France, for the evaporation of Protestant ethnicity in the Irish Republic has points of similarity with the shrinkage of linguistic distinctiveness in peripheral regions like Alsace and Brittany.

The direction in which Northern Ireland is heading is less certain, and hope or despair often seem to depend on how well the last week or the last month has been survived. For this Province, with its reproduction of mutually antagonistic ethnonational blocs from one generation to the next, it is significant that analogies within the EU are elusive. The Basque case has some limited similarity, but a better match lies outside the Union in Bosnia-Herzegovina. The strong religious underpinning to ethnic identity in this former Yugoslav territory makes the parallel a particularly instructive one, as do the roles of neighbouring Serbia and Croatia. Analogies like these are stimulated by the combination of an ethnic interpretation of the Irish situation and an internationally geographical perspective on social structure. They make it analytically important to go in search of the concept of ethnicity in Ireland to augment the more familiar application of the ideas of religion and nationalism. It is true that such parallels are always easy to reject if a naive hope of too much similarity leads to disappointed expectations, but, handled with appropriate discretion, they can give useful insights and, in a case like Northern Ireland, provide a chilling pause for thought.

REFERENCES

Adshead, M. (1996) 'Sea change on the isle of saints and scholars? The 1995 Irish referendum on the introduction of divorce', *Electoral Studies* 15, 1: 138–42.

Akenson, D. H. (1975) *A Mirror to Kathleen's Face: Education in Independent Ireland 1922–1960*, Montreal: McGill-Queen's University Press.

Benson, D. E. and Sites, P. (1992) 'Religious orthodoxy in Northern Ireland: the validation of identities', *Sociological Analysis* 53, 2: 219–28.

Boal, F. W. and Livingstone, D. N. (1984) 'The frontier in the city: ethnonationalism in Belfast', *International Political Science Review* 5, 2: 161–79.

Boal, F. W., Campbell, J. A., and Livingstone, D. N. (1991) 'The Protestant mosaic: a majority of minorities', in P. J. Roche and B. Barton (eds) *The Northern Ireland Question: Myth and Reality*, Aldershot: Avebury.

Boal, F. W., Murray, R. C. and Poole, M. A. (1976) 'Belfast: the urban encapsulation of a national conflict', in S. E. Clarke and J. L. Obler (eds) *Urban Ethnic Conflict: A Comparative Perspective*, Chapel Hill: Institute for Research in Social Science, University of North Carolina.

Bowen, K. (1983) *Protestants in a Catholic State: Ireland's Privileged Minority*, Kingston: McGill-Queen's University Press.

Brewer, J. D. (1992) 'Sectarianism and racism, and their parallels and differences', *Ethnic and Racial Studies* 15, 3: 352–64.

Buckland, P. (1972) *Irish Unionism*, 1, *The Anglo-Irish and the New Ireland 1885–1922*, Dublin: Gill and Macmillan.

Cairns, E. and Mercer, G. W. (1984) 'Social identity in Northern Ireland', *Human Relations* 37, 12: 1095–1102.

Calmy, B. (1987) 'Le long des routes', in M. Sailhan (ed.) *L'Irlande: Les Latins du Nord*, Paris: Autrement Revue.

Cecil, R. (1993) 'The marching season in Northern Ireland: an expression of politico-religious identity', in S. MacDonald (ed.) *Inside European Identities: Ethnography in Western Europe*, Oxford: Berg Publishers.

Census of Ireland 1911: General Report (1913), London: HMSO.

Central Statistics Office (1977) *Census of Population of Ireland 1971*, IX, *Religion*, Dublin: Stationery Office.

—— (1994) *Ireland, Census 91: Summary Population Report*, 2nd series, Dublin: Stationery Office.

Compton, P. A. and Coward, J. (1989) *Fertility and Family Planning in Northern Ireland*, Aldershot: Avebury.

Considère-Charon, M.-C. (1995) 'L'état irlandais et la minorité protestante de 1922 à 1992', *Études Irlandaises* 20, 1: 179–91.

Coulter, C. (1994) 'Class, ethnicity and political identity in Northern Ireland', *Irish Journal of Sociology* 4: 1–26.

Cox, W. H. (1985a) 'The politics of Irish unification in the Irish Republic', *Parliamentary Affairs* 38, 4: 437–54.

—— (1985b) 'Who wants a united Ireland?', *Government and Opposition* 20, 1: 29–47.

Davis, E. E. and Sinnott, R. (1979) *Attitudes in the Republic of Ireland Relevant to the Northern Ireland Problem*, I, *Descriptive Analysis and Some Comparisons With Attitudes in Northern Ireland and Great Britain*, Dublin: Economic and Social Research Institute.

Douglas, J. N. H. (1989) 'Cultural pluralism and political behaviour in Northern Ireland', in C. H. Williams and E. Kofman (eds) *Community Conflict, Partition and Nationalism*, London: Routledge.

Engman, M. (ed.) (1992) *Ethnic Identity in Urban Europe*, Aldershot: Dartmouth Publishing Company.

Fulton, J. (1991) *The Tragedy of Belief: Division, Politics, and Religion in Ireland*, Oxford: Clarendon.

Gallagher, M. (1995) 'How many nations are there in Ireland?', *Ethnic and Racial Studies* 18, 4: 715–39.

Garvin. T. (1988) 'The north and the rest: the politics of the Republic of Ireland', in C. Townshend (ed.) *Consensus in Ireland: Approaches and Recessions*, Oxford: Clarendon.

Girvin, B. (1986) 'Nationalism, democracy, and Irish political culture', in B. Girvin and R. Sturm (eds) *Politics and Society in Contemporary Ireland*, Aldershot: Gower.

Hart, P. (1993) 'Class, community and the Irish Republican Army in Cork, 1917–1923', in P. O'Flanagan and C. G. Buttimer (eds) *Cork: History and Society: Interdisciplinary Essays on the History of an Irish County*, Dublin: Geography Publications.

—— (1996) 'The Protestant experience of revolution in Southern Ireland', in G. Walker and R. English (eds) *Unionism in Modern Ireland*, Dublin: Gill and Macmillan.

Helleiner, J. (1995) 'Gypsies, Celts and tinkers: colonial antecedents of anti-traveller racism in Ireland', *Ethnic and Racial Studies* 18, 3: 532–54.

Kellas, J. G. (1991) *The Politics of Nationalism and Ethnicity*, Basingstoke: Macmillan.

Kennedy, D. (1988) *The Widening Gulf: Northern Attitudes to the Independent Irish State 1919–49*, Belfast: Blackstaff.

Krejci, J. and Velimsky, V. (1981) *Ethnic and Political Nations in Europe*, London: Croom Helm.

Lane, J.-E. and Ersson, S. O. (1987) *Politics and Society in Western Europe*, London: Sage Publications.

Lijphart, A. (1975) 'The Northern Ireland problem: cases, theories, and solutions', *British Journal of Political Science* 5, 1: 83–106.

Lyons, F. S. L. (1967) 'The minority problem in the 26 counties', in F. MacManus (ed.) *The Years of the Great Test*, Cork: Mercier.

MacLaughlin, J. (1996) 'The evolution of anti-Traveller racism in Ireland', *Race and Class* 37, 3: 47–63.

Macourt, M. P. A. (1995) 'Using census data: religion as a key variable in studies of Northern Ireland', *Environment and Planning A* 27, 4: 593–614.

McAllister, I. (1982) 'The Devil, miracles and the afterlife: the political sociology of religion in Northern Ireland', *British Journal of Sociology* 33, 3: 330–47.

McGarry, J. and O'Leary, B. (eds) (1993) *The Politics of Ethnic Conflict Regulation: Case Studies of Protracted Ethnic Conflicts*, London: Routledge.

McKernan, J. (1982) 'Value systems and race relations in Northern Ireland and America', *Ethnic and Racial Studies* 5, 2: 156–74.

McVeigh, R. (1992) 'The specificity of Irish racism', *Race and Class* 33, 4: 31–45.

Mitchell, P. (1995) 'Party competition in an ethnic dual party system', *Ethnic and Racial Studies* 18, 4: 773–96.

Morrow, D. (1995) 'Warranted interference? The Republic of Ireland in the politics of Northern Ireland', *Études Irlandaises* 20, 1: 125–47.

Moxon-Browne, E. (1980) 'Attitudes towards religion in Northern Ireland: some comparisons between Catholics and Protestants', *PACE* 12, 2: 21–3.

Nanton, P. (1992) 'Official statistics and problems of inappropriate ethnic categorisation', *Policy and Politics* 20, 4: 277–85.

Osborne, R. D. (1996) 'Policy dilemmas in Belfast', *Journal of Social Policy* 25, 2: 181–99.

Peach, C., Robinson, V. and Smith, S. (eds) (1981) *Ethnic Segregation in Cities*, London: Croom Helm.

Pringle, D. G. (1985) *One Island, Two Nations? A Political Geographical Analysis of the National Conflict in Ireland*, Letchworth: Research Studies Press.

—— (1989) 'Partition, politics and social conflict', in R. W. G. Carter and A. J. Parker (eds) *Ireland: A Contemporary Geographical Perspective*, London: Routledge.

Ruane, J. and Todd, J. (1991) ' "Why can't you get along with each other?": culture, structure and the Northern Ireland conflict', in E. Hughes (ed.) *Culture and Politics in Northern Ireland 1960–1990*, Milton Keynes: Open University Press.

—— (1992) 'Diversity, division and the middle ground in Northern Ireland', *Irish Political Studies* 7: 73–98.

Samarasinghe, S. W. R. de A. and Coughlan, R. (eds) (1991) *Economic Dimensions of Ethnic Conflict: International Perspectives*, London: Pinter.

See, K. O. (1986) *First World Nationalisms: Class and Ethnic Politics in Northern Ireland and Quebec*, Chicago: University of Chicago Press.

Sillitoe, K. and White, P. H. (1992) 'Ethnic group and the British census: the search for a question', *Journal of the Royal Statistical Society A* 155, 1: 141–63.

Smith, A. D. (1991) *National Identity*, Harmondsworth: Penguin.

Smith, S. J. (1989) *The Politics of 'Race' and Residence: Citizenship, Segregation and White Supremacy in Britain*, Cambridge: Polity Press.

Stephen, F. (1993) 'Integrated education in Northern Ireland: current provision and legislation', in C. Moffat (ed.) *Education Together For a Change: Integrated Education and Community Relations in Northern Ireland*, Belfast: Fortnight Educational Trust.

Sutton, M. (1994) *An Index of Deaths From the Conflict in Ireland 1969–1993*, Belfast: Beyond the Pale Publications.

Wallis, R., Bruce, S. and Taylor, D. (1986) *'No Surrender!': Paisleyism and the Politics of Ethnic Identity in Northern Ireland*, Belfast: Department of Social Studies, Queen's University of Belfast.

Whyte, J. H. (1990) *Interpreting Northern Ireland*, Oxford: Clarendon.

Part III

TERRITORY, NATIONALISM AND THE CONTESTATION OF IDENTITY

INTRODUCTION

In its concern with nationalism and nationalist tropes of identity and place, Part III adopts what is perhaps a more obvious approach to the diversity and fluidity of Irishness. It concentrates on Northern Ireland, but that is where the problems are more intractable, where 'history hurts'. The underlying premises of the three chapters are provided by the conceptualisation of nationalism as an imagined communality, defined in contradistinction to an Other, and also by the idea that in cultural and political ideologies, time is translated into space which then becomes labelled with idealised attributes of different epochs and their functions.

In this latter context, Northern Ireland can be seen as a failure to turn time into space, the result being an incoherent and ethnically fractured conceptualisation of identity. In Chapter 8, Neville Douglas argues that the protection and security of identity in contested territories usually results in violent conflict, whereas social interaction is more commonly defined by competition in societies such as the South where territory and national ideology coincide. He explores the nature of identity and the importance of representations of the Other and Otherness, and reviews the nature of conflicting identity in Northern Ireland. As does Michael Poole in Chapter 7, Douglas finds evidence of new opportunities for politics of accommodation as monolithic depictions of unionism and nationalism break down.

Chapter 9 contains an exploration by Nuala Johnson of the turning of time into space. It demonstrates how what many people regard as being the 'real Ireland' of the West is a product of the processes that transformed the West in general, and the Gaeltacht regions in particular, into repositories of a hegemonic representation of national identity, defined by discourses of premodernism and ethnic purity. This narrative articulated an Irish imagined community in which the West became a synecdoche of Irish identity. It was exclusive and, if initially élitist, capable of appealing to the mass of the population that was not Protestant and/or unionist. The chapter goes on to explore the dichotomy between 'public discourse and political practice', the cultural heartland of the nation developing into a peripheral and economically backward region. Again we can see that the perceived homogeneity of national narrative conceals an axis of diversity – in this case regional inequalities in wealth and opportunity.

Nevertheless, the cultural cement provided by the imagined communality of the West is absent in Northern Ireland. Whereas in the South, irrespective of the dissonance between the mythology of place and the actual life of the people who lived there, the West provided an immensely powerful trope of identity, in the North it is sectarianism that has provided the basis of identity. Chapter 10 explores the failure of unionism to evolve a sense of place that might have legitimated and validated partition. Instead, it was content to define itself largely through the adversarial Otherness of Catholic republicanism. As a more diverse Ireland materialises and identity becomes more open-ended, unionism is left without any imagined communality, lacking any cultural mechanism to translate time into space. The chapter argues, however, that any attempt to formulate the positive cultural iconography necessary to imagine and thereby legitimate a unionist conceptualisation of place inevitably invokes a revised and pluralist Ireland, defined by regional and cultural heterogeneity, notions of hybridity and the equality of rights of citizenship embodied in civic nationalism.

8

POLITICAL STRUCTURES, SOCIAL INTERACTION AND IDENTITY CHANGE IN NORTHERN IRELAND

Neville Douglas

Despite its importance for nation, state and society, identity is a concept which is not easily understood. In the first part of this chapter, I discuss the nature of identity with reference to the following questions: How can identity be recognised? How is it formed? How is it maintained? What is its function? Can identity be changed? I then consider identity in Northern Ireland between 1921 and 1972. This was a period characterised by the politics of dominance and control, a lack of social change and by entrenched individual and group identities. Finally, the chapter will examine how traditional identities have been changed and replaced since the implementation of Direct Rule from Westminster in 1972. Together with protracted sectarian violence, this has removed old political and social certainties and introduced social and economic change and new political structures and challenges in a rapidly modernising world.

THE NATURE OF IDENTITY

In the search for identity I begin, in contrary fashion, with a query not posed above. In responding to the question, 'Who am I?', all individuals begin to place themselves in psychological and social terms, interrogating beliefs and values while defining those ideals that are accepted as right and proper. The question involves the signification of cherished aspirations, acceptable behaviours and right and wrong in moral, social and political contexts. The answers will begin to identify the self and place it within the structures and processes of society (Pred 1985).

Values, beliefs and aspirations are used to identify the self with similarly minded people, a sameness that leads on to the formation of supportive groups and organisations. Thus the self is located in a world of social groups and organisations, ever evolving in a continuous engagement with its environment. The function of identity lies in providing the basis for making choices and facilitating relationships with others while positively reinforcing these choices (Baumeister 1986). In emphasising sameness, group

151

membership provides the basis for supportive social interaction, coherence and consensus. As identity is expressed and experienced through communal membership, awareness will develop of the Other – identities and groups with competing and often conflicting beliefs, values and aspirations. Recognition of Otherness will help reinforce self-identity, but may also lead to distrust, avoidance, exclusion and distancing from groups so defined.

Inevitably, therefore, the social world becomes organised into places of attraction or avoidance: secure places with safe people, or unsafe and even dangerous places with menacing people (Sibley 1995). Consequently, the concept of Otherness limits social interaction beyond our membership groups, reactions to different individuals or groups in social contexts telling us as much about ourselves as it does about them. Clearly, the way we interact with and prefer the company of like-minded people, while having less contact with outsiders, is fundamental to the question of how identity is maintained. The sense of belonging – built upon supportive social interaction – reinforces self-identity while bringing confidence and courage to behaviour. In times of stress or crisis, preservation of self- and group identity becomes the primary concern. In a competing and conflicting world of people and places, social interaction will always prioritise and reinforce established social groups and structures (Kristeva 1991). Thus is identity maintained over time.

Like intelligence, however, identity is not immediately recognisable. For example, the use of colour or race as a defining criterion owes more to subjective prejudice and stereotyping than to impartial analysis (Sibley 1995). Identity is recognised through observation of social interaction and behaviour and is revealed by an individual's views, advocacy of arguments and support for particular groups in different social contexts. Identity will also be expressed in spatial behaviour, as in choice of residential and recreational locations or place of work, such behaviours reflecting the external application of internalised beliefs, values and aspirations. However, the placing of the identity of the Other and the behaviour consequent upon this is not based upon a single foundation. Identity is multifaceted, being built upon, and existing in, a range of human attributes such as language, religion, ethnicity, national feeling and interpretation of the past (Guibernau 1996). Characteristics of gender, age and class can be important, as well as feelings about issues of disability, disadvantage and inequality. Identity may also find expression in art, music or sports and in views on human use of the physical and natural environment. In peaceful, mature societies, identity will find expression in a range of behaviours which will differ in priority and intensity over time and social context.

Having examined how identity is maintained, expressed and recognised, we may now ask how identity is formed and how we come to have specific beliefs, values and aspirations. Put simply, identity is constructed through a learning process at different structural levels of society which is particular to, and continues from, individuals' earliest days within the society into which they are born (Figure 8. 1). At an early age family and kin groups form the context for learning values and social roles. Later, as the child grows, neighbourhood and locality are expressed through play groups, church groups and school attendance. Gradually, extended interaction means that new roles in wider external social frameworks are imposed upon the initial family experience. As children reach young adulthood, awareness of self and family becomes contextualised within the wider world of work, recreation and authority beyond the locality, all against the backdrop of a powerful mass media. Each new element of the social structure expands and conditions individual learning and hence behaviour. Awareness and participation in the wider world means a continued expansion of secondary roles in a complexity of institutional settings. In sum, this is the socialisation process, which provides for ever more complex and subtle storage, recall and transmission of information in individually assessed social and spatial contexts (Fletcher 1976).

Examination of this process of socialisation and social interaction establishes that identity – individual and group – is formed in a similar way in all societies and in all places. How then can it be that, in some cases, similar processes of socialisation result in identities and societies where competitive but generally peaceful social interaction occurs, while protracted violence and social conflict become the norm in others – as in Northern Ireland? The causes of individual and social conflict in any place are invariably found rooted in the depths of historical competition for control and ownership of territory. The past structures the context for contemporary socialisation and behaviour, territorial demands meshing with national aspirations and politics. Nationalism abets the ideological aspiration while politics provides the arena of struggle for power. Control of power confers the means of safeguarding identity and ensuring that the beliefs, values and aspirations emotively, and even mystically, related to place are protected. In contested territories, protection and security of identity usually results in violent conflict. In contrast, where territory and national ideological identities coincide, social interaction is likely to give rise to competition rather than protracted violent conflict (Gottman 1973).

In societies divided by contests over territory and power, the processes of socialisation and interaction usually occur within well-established and separate social structures. Intra-subgroup communication enforces the sameness of identity. In turn, a lack of interaction between subgroups emphasises difference and Otherness. Separate social structures and processes tend to be located in places apart – subgroups live in segregated areas, children are taught separately, and adults work, spend their leisure time, and are even buried apart. Separate social worlds become separate place worlds which in turn become spaces of identity (Soja 1989). Crucially, in these separated societies, isolation encourages the development of stereotypical depictions of

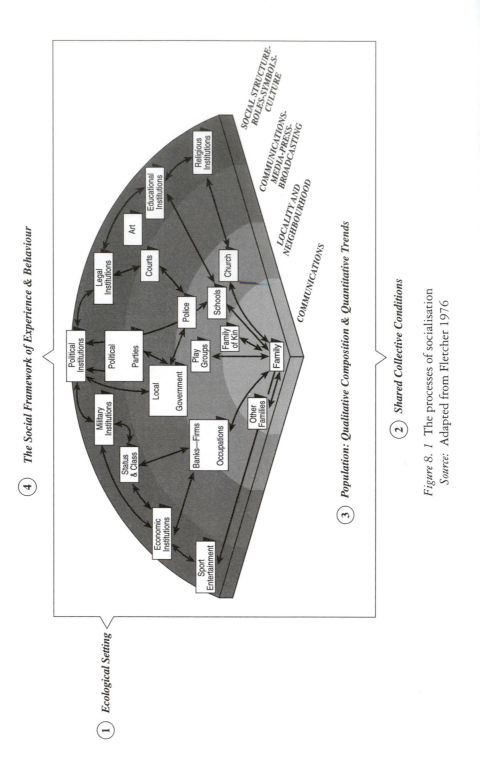

④ *The Social Framework of Experience & Behaviour*

SOCIAL STRUCTURE-
ROLES, SYMBOLS-
CULTURE

COMMUNICATIONS-
MEDIA-PRESS-
BROADCASTING

LOCALITY AND
NEIGHBOURHOOD

COMMUNICATIONS

Religious
Institutions

Educational
Institutions

Art

Legal
Institutions

Courts

Church

Political
Institutions

Political
Parties

Police

Schools

Military
Institutions

Local
Government

Play
Groups

Family
of Kin

Family

Status
& Class

Banks—Firms
Occupations

Other
Families

Economic
Institutions

Sport
Entertainment

① *Ecological Setting*

③ *Population: Qualitative Composition & Quantitative Trends*

② *Shared Collective Conditions*

Figure 8. 1 The processes of socialisation
Source: Adapted from Fletcher 1976

self and the Other, the latter defining all that is offensive, hated and feared. These images can become the prime source of insecurity, leading to psychological as well as spatial distancing and exclusion. The promotion and acceptance of derogatory images of the Other greatly reinforces and supports the positive image of the self, which becomes the stereotype of all that is good, honest, desirable and superior. Divided societies are thus held in their differences by a vicious circle of cause and effect. Existing structures socialise the individual and bequeath traditional roles. In playing these roles the individual reinforces and recreates the established structure. Lack of change, social entrenchment and ossification become the norm. As Dahl (1976: 313) puts it:

> If all cleavages occur along the same lines, if the same people hold the same positions in one dispute after another, then the severity of conflicts is likely to increase. The [person] on the other side is not just an opponent; [she or he] soon becomes an enemy.

In answering questions on the formation, maintenance and description of identity, a functional definition has emerged. Identity is not 'given' in human nature, but is rather learned through social interaction and communication in a complex of social structures, set in specific and distinctive places and epochs. Acceptance of the functionalist or social interactionist perspective of identity is important because it provides not just a means of description but also a method of analysis, which can lead to a deeper understanding of the nature and dynamics of identity. The further analysis of identities in Northern Ireland in this chapter employs such a functional, social interactionist approach. Before moving on to this discussion, however, a final general question about identity must be considered. As identity is formed through socialisation in specific socially structured contexts, is it then possible to change identity by altering social structures and the related social processes? Empirical studies have established clearly that beliefs, values and aspirations are formed early in life and that once in-built – socially and psychologically – they do not change easily (Musgrave 1973). Yet it is also accepted that social and political change, particularly if cumulative over protracted periods of time, will lead to identity change. New social structures provide new ways of expressing identity. Different aspects of socialisation may challenge accepted beliefs and aspirations, as changing interaction evolves into new social groupings in altered contexts. It is indeed possible for divided societies to experience identity change and to shift from conflict with violence towards competition without violence. Having clarified the nature of identity and examined the possibilities for identity change, I now move on to discuss Northern Ireland.

NORTHERN IRELAND IDENTITIES 1921–72

As Chapters 2 and 3 have shown, the partition of Ireland in 1921 superimposed a new unagreed political structure upon a national territory long contested by a multiplicity of identities. In the North, political power and control passed under unionist hegemony, Protestants comprising just over 60 per cent of the total population. In this contested territory, political control facilitated the protection and safeguarding of identity. The beliefs, values and aspirations of unionists could be established authoritatively in the political arena and prioritised in social structures. Conversely, the nationalist Catholic minority (at just under 40 per cent), was made powerless and found its identity threatened. Partition bequeathed new patterns of security and insecurity, fresh possibilities for perceived justice and injustice and increased intensities of sameness and Otherness. Mixing the old sectarian wines in a new political bottle created a process of fermentation leading to the emergence of a deeply divided society. Over the period 1921–72, the range and intensity of sectarian divisions in Northern Ireland increased, the awareness of difference coming to pervade every aspect of life. Distancing and exclusion led to the erection of the Other into a negative and derogatory stereotype (Rose 1971). While strong policies of dominance and control ensured that violent conflict occurred only sporadically during the period, the continued fermentation of sectarian difference slowly but inexorably built up dissident pressure within Northern Ireland.

The identities which formed and entrenched during the politics of dominance and control in Northern Ireland are well described by O'Donnell (1977). Sampling from each community, he asked individuals to choose from a list of adjectives those words which would describe most accurately their own identity, compared to that of the Other. Protestants described themselves, in decreasing order of intensity, as British, loyalist, ordinary, determined, decent people, industrious, conservative and power-holders. Conversely, Catholics described Protestants as power-holders, bigoted, loyalist, Orangemen, British, bitter, ordinary people, brainwashed, determined and murderers. The difference between the strong, sympathetic and supportive self-image and the derogatory cross-group identification is clear. Catholics identified themselves as Irish, long-suffering, ordinary, insecure, decent, deprived, unfortunate, fine people, nationalistic and reasonable. Protestants identified Catholics as ordinary people, Irish, priest-ridden, 'breed like rabbits', republican, bitter, superstitious, brainwashed, nationalistic and not bad. Unlike the Protestant self-image, which is strong and confident, the Catholic self-image is that of an insecure, deprived, powerless people whose aspirations are unattainable. Protestant identification of Catholics is derogatory, but also patronising of a people who have been led astray.

The functioning of unionist-Protestant identity

In its first piece of major legislation, the Local Government (NI) Act of 1922, the unionist government began to mould a new political structure and stamp its authority upon the region. In 1920, prior to partition, a proportional representation (PR) voting system had been introduced for local government elections in Ireland. Employed in the 1920 local elections, this gave greater representation to nationalist parties in the North and to Labour in Belfast. The 1922 legislation removed the PR system, replacing it with the old single-seat constituency plurality system. It also allowed for the reconsideration and possible redrawing of electoral boundaries within the existing local administrative areas. This Act entrenched unionist power at local level and complemented regional government control – Northern Ireland became a unionist territory for a unionist people. In this case and all others, the maintenance of the constitutional link with Britain was the priority of all unionist government policy. Control and security came before recognition of minority rights while unionist domination of local government became so locked in by the electoral system that many seats were seldom contested (Table 8. 1).

Table 8. 1 Northern Ireland: percentage of County Council and County Borough seats uncontested in eight local government elections between 1946 and 1967

County Councils	Percentage seats uncontested	County Boroughs	Percentage seats uncontested
Antrim	86	Belfast	25
Armagh	85	Londonderry	83
Down	94		
Fermanagh	95		
Londonderry	96		
Tyrone	99		

Source: Adapted from Lawrence (1975)

Furthermore, the politics of control reached into the state apparatus so that applications for posts in the Northern Ireland Civil Service were scrutinised most carefully to ensure the success of Protestant candidates. As Patrick Shea, one of the few Catholics to reach the higher levels of bureaucracy, writes in his autobiography (1981: 197): 'I sometimes felt like a cuckoo in the nest. Protestants would ask me how I "got in". And

Catholics, particularly in Belfast, looked on me with suspicion. I must have influence; I was probably a "bad Catholic", who had gone over to the other side.' Education policy produced a segregated structure of primary and secondary schooling (Birrell and Murie 1980), while control of public-sector housing allocation by local councils gave unionists the advantage in securing homes. Employment practices at local council level followed the lead of regional government. Mirroring the public sector, informal social and economic structures also emerged in the private sector to make more complete the social, psychological and spatial divisions in Northern Ireland society. The meaning of these divisions is described with great clarity by Robert Harbinson (1960: 16) as he recalls his childhood in Belfast during the 1930s. Living in the 'Village' area, he could see across the low-lying Bog Meadows and west Belfast to the black basalt mountain beyond:

> In terms of miles the mountain was not far, and I had always longed to explore it. . . . But the mountain was inaccessible because to reach it we had to cross territory held by the Mickeys [Catholics]. Being children of the staunch Protestant quarter, to go near the Catholic idolaters was more than we dared, for fear of having one of our members cut off.

As such examples of unionist-Protestant behaviour illustrate, identity came to be expressed through dominance and control. Social and spatial distancing became particularly important as it ensured the maintenance and recreation of identity in each new generation. And yet, as Harbinson shows, separation was accompanied by proximity, by what Stewart (1977) has famously described as the narrow ground, and the constant presence of threat. Despite O'Donnell's finding of a strong Protestant self-identity, the constant close presence of the Other, and its greater natural population increase, created an insecurity among the unionist population. All too often overlooked, this neurosis is an important facet of unionist identity. Partition was interpreted by northern unionists as a betrayal of their rightful British heritage, because it partially detached their Britishness and created them a minority within the island of Ireland. The depth of the perceived betrayal was well put by Rudyard Kipling in his poem, 'Ulster 1912':

> The dark eleventh hour
> Draws on and sees us sold
> To every evil power
> We fought against of old.
> Rebellion, rapine, hate,
> Oppression, wrong and greed
> Are loosed to rule our fate
> By England's act and deed.
> Before an Empire's eyes
> the traitor claims his price,

What need of further lies
We are the sacrifice.

The sense of betrayal is clear but is followed by defiance in the last four lines of the poem:

What answer from the North?
One Law, one Land, one Throne.
If England drive us forth
We shall not fall alone.

(Kipling 1919: 10–12)

It is only by recognising this sense of betrayal, matched with defiance and an utter determination not to lose control of their destiny, that the translation of identity into action by unionists between 1921 and 1972 can be understood. Indeed, the behaviours of controlling, distancing and excluding became in themselves distinct values and aspirations, not as means to an end, but as facets of identity in their own right to be placed alongside the deeper ideas of right and wrong long planted in the collective unionist consciousness.

The functioning of nationalist-Catholic identity

Paralleling the unionist experience, partition left nationalists in the north-east of Ireland with a deep sense of betrayal and injustice. From belonging to a majority seeking political independence for a natural, even God-given, unit of territory, they were transformed into a minority within an artificially constructed political unit. Most difficult of all to accept was the realisation that they had been abandoned by their fellow Irish nationalists. The biased outcomes of the Local Government Act after 1922 confirmed the worst suspicions as to their future role in the territory of the new unionist regime.

Powerless in this developing unionist hegemony, new aspects of identity and behaviour began to augment the traditional values of the nationalist axis. Negative self-images of a deprived and rejected community were soon being counteracted by tactics of defiance and non-recognition of the new state and abstention from involvement in its structures and processes. These strategies became the means of protecting and maintaining traditional values and cherished aspirations. Thus in 1921–2, Counties Tyrone and Fermanagh, together with ten Urban Districts and six Rural Districts – mostly in areas adjacent to the new international boundary – refused to recognise the Northern Ireland State and pledged allegiance to Dáil Éireann. Non-recognition of the state was further expressed in a continuing refusal to vote in elections. In the 1924 local government elections, widespread abstention handed control of almost all councils, regardless of voting system or local majorities, to unionists. The exceptions were seven Urban Districts and one Rural District where nationalists did vote in sufficient numbers to gain control. The outcome of this election became a fundamental generator of nationalist identity during the following decades. The overwhelming nature of the unionist victory was attributed to the implementation of the 1922 Local Government Act and the perceived widespread gerrymander that it engineered. Belief in this device as a method of control caused the nationalist community to self-identify as a deprived people, while perceiving the unionists as illegal power-holders (Douglas 1982), a rendition which became a vital factor in the formation of stereotypes of Otherness in both communities. Abstention from voting remained a significant strategy well into the 1930s. Even after participation did become general, successful nationalist candidates often refused to take their seats. It was not until 1965 that the Nationalist Party took on the mantle of official opposition in the regional parliament at Stormont.

Defiant non-involvement in political structures and rejection of citizenship were soon augmented by an alternative and opposing set of social and economic nationalist structures, often generated by, and focused on, the Catholic church. Separate educational and teacher-training structures resulted as much from Catholic preferences as they did from unionist legislation. In the setting up of the National Health Service in 1947, for example, all hospitals were placed under a single Hospitals Authority. However, the Mater Hospital in Belfast remained outside the system, being seen as part of the healing ministry of the Catholic church with its own chapel and its own Catholic nursing sisters (Barritt and Carter 1962). Maintenance of separate educational, teacher-training and health structures put a heavy financial cost on the nationalist community and particularly upon the Catholic church. It was the price paid, however, to sustain rejection of the state and protect nationalist identity, further protected and strengthened, moreover, by financial structures such as the Credit Unions, charities (most notably, St Vincent de Paul) and sporting and cultural organisations such as the GAA (Gaelic Athletic Association). All these aspects of the practical functioning of identity were supported by a separate nationalist press, which highlighted the biased sectarian repression of the unionist state in its own partisan fashion.

Nationalist identity between 1921 and 1972 was expressed functionally through a rejection of, and abstention from, the unionist regime and by the positive development of these alternative social and cultural structures, which assumed their own political significance. Like unionists, the nationalist community sought distancing, exclusion and physical separation, behaviours which became not just expressions of sameness but integral values and aspirations of identity in their own right. For both communities, behaviour became identity and identity behaviour. Over five decades, the two segments of Northern Ireland society, each with their separate internal structures and processes, preserved and protected their sameness, while lack of contact with the Other entrenched strong negative images and derogatory

160

stereotypes. Lines from W. B. Yeats's poem, 'Meditation in Time of Civil War and Rebellion', reflecting on the earlier Irish Civil War, aptly describe the outcome of sectarian division in Northern Ireland:

> We are closed in, and the key is turned on our uncertainty. . . .
> We had fed the heart on fantasies,
> The heart's grown brutal from the fare;
> More substance in our enmities
> Than in our love.

(Yeats 1933: 232)

Yet – despite all – community segregation was never complete over these decades. Individuals and groups continued to emphasise other values and aspirations and reject identity founded solely on facets of nationality and religion. Problems of unemployment, poverty and inequality affected both unionist and nationalist working classes. The Belfast Labour Party, and later the Northern Ireland Labour Party, concentrated on such problems, blaming them on the fundamentally capitalist ethos of the unionist government. Their unwavering argument was that sectarianism divided workers and facilitated unhindered exploitation to the benefit of the ruling capitalist class (Pringle 1985). In periods of great economic hardship, as in the early 1930s, or wider social upheaval such as that which followed the 1945 British General Election, trade unions and socialist labour parties gained wide support from the working-class communities in Northern Ireland. But support could never be maintained as unionist politicians spread dire warnings of the threat to the constitutional link with Britain and, encouraged by those same unionist leaders, nationalists retrenched to traditional strategies of defiance and non-recognition.

In rural areas, too, separation was far from complete in day-to-day living. As Harris (1972) points out, relations between Catholics and Protestants were for the most part peaceful, with good neighbourliness and co-operative rural work practices existing between individual farmers of different religions. Considerable efforts were made to avoid insult or any display of hostility to the Other, usually through careful use of language and by avoiding sensitive political and religious topics. This tolerant acceptance of prejudice in the rural context is well described by Seamus Heaney (1995: 194), when he writes about the Protestant farmer standing outside in his Catholic neighbour's yard at night and not going into the house until he heard the family finish its prayers. This action 'was not fundamentally intended as a contribution to better community relations . . . [nor a] social obligation' but instead 'a moment of achieved grace between people with different allegiances'. In the rural context, as in the urban, such small harmonies existed among the dissonance between conflicting Others.

In summary, three observations can be made about identity and its functioning in Northern Ireland between 1921 and 1972. First, identities were formed and became set fast in a mould, or rather in two separate moulds, which changed little over the five decades. Lack of change and the existence of deep divisions were accepted as inevitable facts of life. Even global events, such as World War II, had little real impact upon sectarian mind-sets or the relationships between them. It was only in the 1960s that the first waves of change and modernisation elsewhere in the world began to erode the bastioned edges of Northern Ireland society. The second point concerns the tightly knit and integrated nature of both unionist and nationalist identities. Within the small region that is Northern Ireland, the memberships of each group, whether leaders or followers, were often educated together and related through marriage; urban migration brought together town and country and employment was usually found through personal contacts. Group members worshipped together, played and watched games together, worked together and resided together in their own areas. Each community shared a rapid and universal spread of supportive information, provided by its own sectarian press. For both unionists and nationalists, living and life did not appear limited and individual and family hardships were counterbalanced by these feelings of group security. Finally, both unionist and nationalist identities – channelled through socialisation within the limiting sectarian context – became based upon and expressed through a deliberately limited number of facets related to religion and the constitution but little else. So was the key turned on uncertainty, diversity and shades of opinion. Such a divided society could hardly change itself from within nor could it be transformed easily by outside forces.

NORTHERN IRELAND IDENTITIES 1972–96

The removal of the regional parliament at Stormont in 1972 and the imposition of Direct Rule from Westminster marked a psychological, social and political watershed of fundamental significance in the history of Northern Ireland. Direct Rule, orchestrated from the Northern Ireland Office, set out along the road to reform with the long-term overarching aim of replacing the politics of control and dominance with those of accommodation. It is difficult to overestimate the significance of this external force in bringing about social and political change through the creation of new structures and processes during the period since 1972. These aimed at facilitating equality of access for both segments of society to the policy-making and policy-administering arms of government. Both British government and the Northern Ireland Office believed policies of accommodation to be the only means of ending violent social conflict and providing an arena structured for peaceful competition.

The psychological effects of political upheaval on historic identities were profound. The unionist community ceased to self-identify as power-holders and the positive, self-confident facets of their identity were replaced by doubt and suspicions of conspiracy by those ranged against them. Like the nationalist community in 1921, unionist identity began to take on facets of powerlessness, deprivation and insecurity. For the nationalist community, conversely, the political upheaval had much less significance in terms of identity. Although the removal of the Stormont parliament, regarded as the source of their suppression for five decades, was a cause for great rejoicing, reservations remained. It still had to be demonstrated that British control would or could give due recognition to nationalists' fundamentally Irish identity and its inherent aspirations. In the years immediately following 1972, Northern Ireland slipped into a cauldron of uncertainty, in which new structures engendered intense and violent conflict as the old unchanging certainties and identities were threatened.

Structural change

The first steps in structural change under Direct Rule were distinctly unpromising as the 1974 power-sharing Executive failed ignominiously, destroyed by efficiently organised and defiant loyalist opposition. Swiftly learning the lesson that grand plans for rapid change are a recipe for violence in deeply divided societies, the Northern Ireland Office set about implementing a slow policy of incremental change. In time, this strategy produced a sturdy raft of socio-political policies and implementing agencies on which new structures were built and new processes of social interaction developed.

No fewer than seven significant governmental axes of accommodationist transformation can be identified.

1 The first enduring structural change occurred in 1972–3 with the implementation of a new local government pattern of twenty-six District Councils, elected by proportional representation. In the first elections held under this system in 1973 every one of the 526 seats was contested, with a 68 per cent turnout of voters. Clearly the new structure for local political decision-making was accepted as fair and proportional and – as such – provided unbiased access to power. An arena had been created into which both communities could enter on an equal footing.

2 Employment practices in both public and private sectors had been characterised by discrimination during the pre-1972 period of unionist control. The illegality of such customs was established by the Fair Employment Acts of 1976 and 1989, which established in law the requirement of equal opportunity and fair employment practices in all employment sectors in Northern Ireland. A Fair Employment Commission, endowed with strong legal powers, was set up to actively monitor employment procedures.

3 Segregated education at primary and secondary levels was clearly a

fundamental force in the socialisation of difference and division. A policy for integrated education was introduced in the 1980s, enabling the Department of Education for Northern Ireland to facilitate the establishment and funding of totally integrated schools. This policy recognised and promoted a grass-roots integrated movement that, in the early 1980s, began to provide schools in which children from the segregated communities could be educated together. The ethos of integrated education was strengthened by the introduction of a core curriculum in 1987, which included a component concerned with 'Education for Mutual Understanding', aimed at encouraging cross-community school contacts. A 'Cultural Heritage' component also ensures that children in segregated schools learn about the region's past through a common history curriculum.

4 To reinforce the accommodatory thrust of these policies, a Community Relations Council, established in 1987 with government funding, was charged with creating structures, especially at a local scale, through which safe and supportive cross-community contact and social interaction could occur.

5 The Northern Ireland Housing Executive, set up and funded by government, was required to provide public-sector housing throughout Northern Ireland solely on the basis of need.

6 The Industrial Development Board (IDB) and the Local Economic Development Unit (LEDU) were established by government with the respective aims of attracting inward investment and industry and facilitating local industrial enterprise.

7 The Northern Ireland Office of the Commissioner for Complaints (the Ombudsman) was established to conduct independent investigations and make legally binding rulings on complaints of discrimination, brought by any individual, irrespective of religion or national aspiration.

It must also be emphasised that transformation has not been confined to government policies alone. As often occurs after policy formation and implementation, unforeseen changes cause considerable social upheaval. Such was the outcome of the implementation in Northern Ireland of government health policies concerned with hospital rationalisation and moves to care in the community. The changes in practice and provision of health care from the 1980s to the present have caused a reactive and massive burgeoning of special-issue, caring and pressure groups within the voluntary and charitable sector. At present over 250 structured groups are affiliated to the umbrella organisation, the Northern Ireland Council for Voluntary Action (NICVA). These are concerned, for example, with issues of illness, ageing, dying, disability, homelessness and poverty. In pursuing their objectives, these issue-related agencies pay scant regard to traditional sectarian divisions.

Wider forces for modernisation

This raft of public and private structures of transformation did not result simply from forces within Northern Ireland working in independent isolation. Wider social forces, continental and even global in extent, have impinged upon, and given substance and impetus to, the changes evolving in the small regional unit of Northern Ireland. Chief among these is the globalisation of information through electronic media, a process which particularly in the past decade has begun to produce a wider imagined international community. The European Union (EU), for example, is promoting a Europe-wide media system as a way of creating a more homogeneous community conscious of shared traditions (Morley and Robins 1995). As a peripheral region, Northern Ireland is also a beneficiary of extensive EU funds, financial inputs which have widened the awareness and the priorities of many occupational groups in business and agriculture. The tensions placed upon traditional identities and loyalties by Europeanisation were well exemplified during the EU ban on British beef, caused by the incidence of BSE (bovine spongiform encephalopathy). The strong farming lobby in Northern Ireland demanded that the European Commission treated the Province separately from the rest of the UK, and on a par with the Republic of Ireland, because of the much lower incidence of the disease and better systems of recording cattle herds. Simultaneously, many of these farmers, as unionists, castigated the British government for including the Irish government in talks on the future of Northern Ireland.

The general western decline in church attendance and religious observance is another aspect of modernisation which has been particularly prevalent among the young, many of whom now see little religious relevance in Protestant or Catholic identities or in the established churches (Livingstone *et al.* 1996). The expansion of third-level education, which spreads new ideas in new places, has also provided an important arena for challenging traditional identities and change. More generally, the 'rolling back of the state' in the UK, together with the privatisation and semi-detaching of many public utilities, has created a much more questioning, suspicious and involved civil society, increasingly sceptical of the rhetoric of political leaders. Northern Ireland has not been immune from this modern trend.

Thus opportunities for expressing citizenship, made possible by the new accommodationist structural framework and the wider forces of modernisation, exist side-by-side with brutal, violent conflict. To an increasing number of citizens, conflict is unjustified and is judged to be self-defeating. Strong counter-insurgency measures by the British Army have also been incapable of producing 'victory'. In this environment of no winners, augmented by modernisation and structural change, an increasing war-weariness has emerged, accompanied by a questioning of old identities and traditional methods of achieving aspirations.

Interaction and social change

The forces of modernisation impinging from outside, and the new social and political structures built internally through the politics of accommodation, have been and continue to create a new 'place-world' in Northern Ireland. A wide range of supportive and interdependent cross-community social interaction has manifested itself in many different ways. In order to explore further the implications for social change of this combination of restructuring and wider forces of modernisation, I will develop four of the axes and arenas of transformation cited above: local government; fair employment legislation; education; the voluntary sector.

First, in the reformed local government arena, cross-party agreements and forms of consociational power-sharing have evolved. In 1989, only seven of the twenty-six councils were controlled by a single party, while, in 1992, fifteen agreed to share and rotate the office of mayor on a six-monthly basis between political parties of different persuasions. The first to do so was Dungannon Council, traditionally divided and with a finely balanced sectarian demography. In March 1992, it declared its area a 'violence free zone', at the same time calling upon all paramilitaries to recognise the rejection of violence by the people of Dungannon. Again, Derry District Council elected a Democratic Unionist Party (DUP) councillor as Mayor of the city in 1992. This was made possible through the support of fourteen members of the Social Democratic and Labour Party (SDLP), the largest nationalist party in the thirty-member council. Such moves towards cross-community sharing were of considerable symbolic significance in a city where, prior to 1972, unionists had maintained control through a notorious gerrymander (Curran 1946; Douglas 1982).

Second, the District Councils provide one example of an arena in which the application of fair employment legislation is actively removing discrimination in the workplace. By 1993, the Fair Employment Agency had published reports on employment practices in seventeen of the twenty-six districts. In eleven cases, there was no discrimination by religious affiliation in the manual workforce. Of the six remaining councils, three showed evidence of discrimination against Catholics and three against Protestants. All councils now employ Community Relations Officers, who (in conjunction with the Community Relations Council, which helps with advice and funding) work to encourage cross-community programmes which develop social and cultural interaction. At present, councils are involved in developing 'District Partnerships' in which significant finance is available from the EU for programmes of employment, urban and rural regeneration and social inclusion (NICVA 1996). Thus, most local councils now provide at least a potential forum for cross-community social inclusion and supportive interaction. However, it is very difficult to assess the significance of this social interaction in terms of identity formation and change and to establish the extent to which it is financially led tokenism. Certainly, forms of triumphalism and remnants of traditional unionist control and exclusive identities are still to be found in the eastern Protestant heartland. More positively, the Community Relations Council, as its Annual Reports show, not only supports local councils in their integrative role but has helped in the creation and funding of a great number of widely dispersed social and educational cross-community groups working for reconciliation through social inclusion.

The Fair Employment Commission has been active since 1976 in establishing employment practices, which safeguard equal opportunity for all sectors of society. By ensuring that employment is based upon 'merit' for those born within Northern Ireland, there has been increased Catholic representation in the workplace and consequent cross-community interaction, particularly in the public sector. However, although the third survey of social attitudes (Gallagher 1993) found that 60 per cent of Protestants and 88 per cent of Catholics were in favour of the religious monitoring of workforces, the outcomes of this policy in terms of accommodation are suspect. Clearly, 'merit' is a concept which is very difficult to assess objectively and when it is complicated by politically engineered 'goals, targets and timetables', the social change which results may not be seen in a positive and supportive context. Some unionists see the policy as a form of positive discrimination which puts them at a disadvantage and further undermines the roots of their identity.

Third, Northern Ireland now has thirty-five integrated schools, educating just under 2 per cent of the 340,000 children of school age. This small percentage is likely to grow as, despite lukewarm support from some churches and strong opposition from others, 67 per cent of Protestants and 74 per cent of Catholics support greater mixing in primary schools (Gallagher 1993). In terms of identity formation, an alternative now exists to the traditional system of education. The expansion of third-level university education since 1972 has also had important effects upon the processes of socialisation and social change, the most striking manifestation being the emergence of a Catholic middle-class meritocracy. The effects of newly acquired bourgeois aspirations, often superimposed on working-class roots, have been twofold. As discussed in Chapter 5, this group, having found employment and careers in the Civil Service, semi-detached government agencies, the legal profession and teaching, has acquired a stake in the political and economic system of Northern Ireland. The expression of citizenship in a place-world of diverse social interactions has caused, at very least, the questioning of identity. Again, new aspirations and prosperity have resulted in residential mobility away from traditional group heartlands. A much more diversified and confident community has emerged in which, to some degree, economic and social issues have been separated from the constitutional imperative (O'Connor 1993). Third-level education appears to have had a much less positive effect upon the unionist population. The increasing number of young people from this group moving elsewhere in the UK for university education has had a perceived adverse effect upon the maintenance of unionist identity, particularly as only two students out of every ten who move out return permanently (Osborne and Miller 1987). This is but one facet of the relatively less secure nature of unionist identity.

Finally, cross-community social interaction, within the many new structures of the voluntary and charitable sector, has led to widespread supportive accommodation. In coming together to campaign against health and social service policies and to push for the needs of the disabled and disadvantaged, voluntary and issue groups are creating precisely the supportive context in which, as Allport (1954) argues, intolerance, prejudice and stereotypes are broken down. The potential social interaction created by these groups can be exemplified by the work of the 'Save our Hospital' campaigns in many towns, including Downpatrick, Dungannon, Ballymoney, Enniskillen, Larne, Omagh and Ballycastle. Coming together to work for local perceptions of the greater good has been a new departure in such places. As Dahl (1976: 313) states: 'Those who are different but who come together to form alliances which are useful in the present and may be needed in the future are less likely to become enemies'.

Contemporary identities in Northern Ireland

Thus it can be argued that the Northern Ireland socio-political arena, restructured over the period since 1972 by the policies of accommodation and forces of modernisation, has removed many of the barriers to social change. There now exists a range of cross-community social interactions and methods of communication which weave through the traditional boundaries of sectarian division. Socialisation and identity formation now take place in a more diverse and flexible 'place-world'. Traditional structures and values remain strong but now sit in a wider and more challenging socio-political context.

If this argument is a valid one, Northern Ireland society can no longer be described or adequately explained as a deeply divided, two-segment plural society in which identities are exclusively unionist or nationalist. A society in transition now exists in which, most significantly, there has been a decline in the blind acceptance of derogatory stereotypes. Today the question 'Who am I?' evokes answers of greater diversity, complexity and subtle caveats among many citizens who are prepared to accept that 'good, tolerant and acceptable' may define the Other as well as 'bad, bigoted and unacceptable'.

Thus I argue that society in Northern Ireland now comprises two ideological poles, one occupied by a segment of the unionist population and the other by the republican section of the nationalist axis (Figure 8. 2). The identity of the 'polar' unionists remains singularly prioritised upon the constitutional position of Northern Ireland within the United Kingdom. This identity has strong facets of insecurity, isolation and defiance. It retains stereotyped sectarian images, perceives conspiracies against it from all sides and is stubbornly inflexible. It is willing to justify violence to protect and maintain its fundamental values. The republican 'polar' identity is more confident and assured but remains narrowly based on the fundamental value and aspiration of the reunification of Ireland and the removal of all things perceived as British. In common with the unionist antithesis, it is willing to use violence to further its aims.

In contrast, the accommodatory unionist and nationalist sectors of society are to be found closer to the centre of the spectrum. These support good cross-community relations which are built upon a range of values and aspirations, different facets of which vary in importance in time and place. They are not identities which constantly prioritise constitutional imperatives. As Figure 8. 2A shows – supporting the argument in Chapter 7 that there are many attributes to ethnicity – it is better not to view centrist identities as forming a single part of society. They should be thought of as two segments which retain separate national and cultural facets of identity, but which converge in mind and in behaviour on certain questions and in particular situations. Figure 8. 2B suggests some of the issues on which centrist segments can, and indeed have, become a single group whose common behaviour is driven by shared values and aspirations. Centrist unionist identity accepts the justice of equality of opportunity in all aspects of economic and social life, irrespective of religious or national affiliation, and also the necessity of some form of power-sharing in any regional assembly set up to run the affairs of Northern Ireland. Centrist nationalist identity puts equality or parity of esteem, treatment and opportunity ahead of the early reunification of Ireland. The maturity of centrist groups is seen in their abilities to hold to and act upon a diversity of facets of identity, which are normally considered contradictory in a divided society. Thus centrist unionists, while desiring to maintain the union, have advocated closer transport and economic links with the Irish Republic as well as treatment for agriculture in Northern Ireland comparable to that in the Irish Republic rather than the United Kingdom. Centrist nationalists, while maintaining the aspiration of a united Ireland, are also involved in improving economic and social structures within Northern Ireland which will decrease the likelihood of reunification. In all these behaviours, an informal 'peace process' is taking place on the ground. New ideas and projects create local coalitions and social interactions which consciously avoid zero-sum politics and encourage the growth of a more reconciled civil society.

There are other issues, however, which, when they arise, push groups with centrist identities towards the poles (see Figure 8. 2C). Constitutional issues such as the suggestion of joint Irish/British control in the administration of

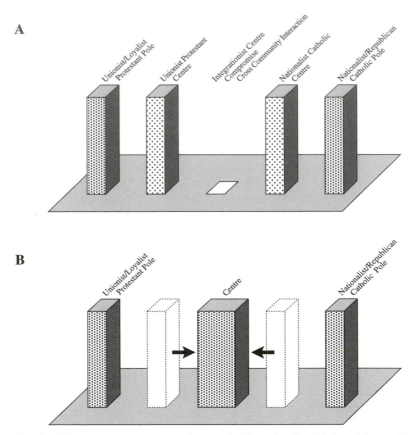

Situation 1- *Communities' reaction to social issues - health care (e.g. hospital closure), drug problems, disability and disadvantage, extreme terrorist violence.*

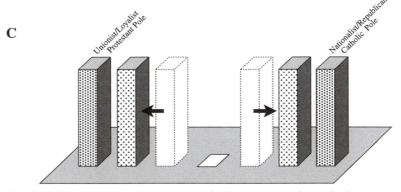

Situation 2- *Communities' reaction to constitutional and security issues - future role of government of Irish Republic in Northern Ireland, politics e.g. joint authority, future role, name and structure of Royal Ulster Constabulary*

Figure 8. 2 The nature of identity in Northern Ireland

170

Northern Ireland send most centrist unionists sliding towards the position of their polar co-religionists. Equally, security issues, traditional sectarian marches and lack of change in the nature and name of the Royal Ulster Constabulary push centrist nationalists away from their integrationist and conciliatory position.

CONCLUSION

In transitional societies gradually emerging from deep sectarian divisions it is always easier for those at the extreme ideological poles, with their willingness to use insult and violence, to resurrect sectarian tension and so block the path to change. In their desire to maintain traditional divisions, the extremist poles are mutually interdependent. Vicious reactions from one pole to violence and insult generated by the Other provides the lifeblood for their continued existence. The refusal by centrist groups to react to violent or derogatory behaviour poses a fundamental threat to both poles. The issue of continuing supposedly long-established Orange Order marches through once-Protestant areas which have become Catholic through residential change is a contemporary example. The violence generated by the unionist pole in response to the police refusal to allow the Orange Order to march from Drumcree Church to the centre of Portadown by way of the nationalist Garvaghy Road in July 1996, and the reactive violence by nationalists when the original decision was reversed to allow the march, both stemmed from a sense of insecurity and fear. Unionists feared a loss of traditional rights and the removal of their constitutional position by incremental change, inherent in the so-called peace process. Nationalist violence arose from fear of a regression back to the former politics of control and the loss of rights and recognition gained since 1972.

The roots of this violence remain strong despite the environment of social and political transition. It is at such times of crisis that the strength, resilience and maturity of the accommodatory centre is most seriously tested. Sectarian violence is anathema to the centrist identity, which has to be sufficiently strong to resist the call to return to old ideologies and traditions. The voice of reason at times of orchestrated extremist violence is lost amidst the mayhem in which the burning of buses creates a far greater psychological impact, and gains more media attention, than the buttressing of cross-community bridges. It may even seem that the centre has disappeared, but it has not. Its strategy at times of extreme violence is to lie low, hold onto beliefs, values and aspirations and guard the bridges which may be used again when extremist hatreds have been sated. Structural–functional analysis of identities in Northern Ireland confirms social change and a slow shift away from entrenched, narrowly classified, exclusive and stereotyped identities. New social structures have influenced the processes of socialisation and the greater diversity of social behaviour has led to more subtle definitions of self and the Other. However, these refined definitions of sameness do not exist free of challenge and tension. They sit alongside and intermingle with traditional, largely unchanged identities, which remain based upon derogatory stereotypical images of Otherness. Discriminatory behaviour is aimed equally at centrist elements in their own community who are willing to work for change. However, it is argued here that tensions within the traditional communities in Northern Ireland are, paradoxically, sure signs of social change. The shift towards reconciliation, tolerance and acceptance of difference is slow and halting and small incremental steps are no more than can be expected. As yet, no acceptable and peaceful way has been found for nationalists to express fully their Irishness and, at the same time, for unionists to express fully their Britishness. Yet there is no doubt that society in Northern Ireland is in transition and that there exists now a society and a place unimaginable in the 1960s.

A quarter of a century is a brief span in the evolution of a deeply traditional and divided regional society, yet much has been achieved through the structural development and functional implementation of accommodatory policies. Despite violent opposition from the extreme ideological poles, the journey of transition along the rocky road to the acceptance of diverse identities in a common territory can continue to bear fruit, but only while the politics of accommodation remain – and are seen to be – implemented with impartial even-handedness.

REFERENCES

Allport, G. W. (1954) *The Nature of Prejudice*, Cambridge, MA: Addison Wesley.

Barritt, D. P. and Carter, C. F. (1962) *The Northern Ireland Problem*, London: Oxford University Press.

Baumeister, R. (1986) *Identity: Cultural Change and the Struggle for Self*, Oxford: Oxford University Press.

Birrell, D. and Murie, A. (1980) *Policy and Government in Northern Ireland: Lessons in Devolution*, Dublin: Gill and Macmillan.

Curran, F. (1946) *Ireland's Fascist City*, Derry: Derry Journal.

Dahl, R. A. (1976) *Pluralist Democracy in the United States*, Chicago: Rand McNally.

Douglas, J. N. H. (1982) 'Northern Ireland: spatial frameworks and community relations', in F. W. Boal and J. N. H. Douglas (eds) *Integration and Division: Geographical Perspectives on the Northern Ireland Problem*, London: Academic Press.

Fletcher, R. (1976) *The Framework of Society*, Milton Keynes: Open University Press.

Gallagher, A. M. (1993) 'Community Relations', in F. Stringer and G. Robinson (eds) *Social Attitudes in Northern Ireland: The Third Report*, Belfast: Blackstaff.

Gottman, J. (1973) *The Significance of Territory*, Charlottesville: University Press of Virginia.

Guibernau, M. (1996) *Nationalisms: The Nation State and Nationalism in the Twentieth Century*, Oxford: Polity.

Harbinson, R. (1960) *No Surrender: An Ulster Childhood*, London: Faber.

Harris, R. (1972) *Prejudice and Tolerance in Ulster*, Manchester: Manchester University Press.

Heaney, S. (1995) *The Redress of Poetry*, London: Faber.

Kipling, R. (1919) *The Years Between*, London: Methuen.

Kristeva, J. (1991) *Strangers to Ourselves*, 2nd ed., New York: Columbia University Press.

Lawrence, R. J. (1975) *The Northern Ireland General Elections of 1973*, Belfast: HMSO.

Livingstone, D. N., Boal, F. W. and Keane, M. (1996) *Them and Us? A Survey of Catholic and Protestant Churchgoers in Belfast*, Belfast: CCRU.

Morley, D. and Robins, K. (1995) *Spaces of Identity: Global Media, Electronic Landscapes and Cultural Boundaries*, London: Routledge.

Musgrave, P. W. (1973) *The Sociology of Education*, London: Methuen.

NICVA (1996) *Partners for Progress: The Voluntary and Community Sector's Contribution to Partnership Building*, Belfast: NICVA.

O'Connor, F. (1993) *In Search of a State: Catholics in Northern Ireland*, Belfast: Blackstaff.

O'Donnell, E. E. (1977) *Northern Irish Stereotypes*, Dublin: College of Industrial Relations.

Osborne, R. D. and Miller, R. L. (eds) (1987) *Education and Policy in Northern Ireland*, Belfast: Queen's University and University of Ulster Policy Research Institute.

Pred, A. (1985) 'The social becomes the spatial, the spatial becomes the social: enclosures, social change and the becoming of places in the Swedish province of Skane', in D. Gregory and J. Urry (eds) *Social Relations and Spatial Structures*, London: Macmillan.

Pringle, D. (1985) *One Island, Two Nations? A Political Geographical Analysis of the National Conflict in Ireland*, Letchworth: Research Studies Press.

Rose, R. (1971) *Governing Without Consensus*, London: Faber.

Shea, P. (1981) *Voices and the Sound of Drums: An Irish Autobiography*, Belfast: Blackstaff.

Sibley, D. (1995) *Geographies of Exclusion*, London: Routledge.

Soja, E. W. (1989) *Postmodern Geographies: The Reassertion of Space in Critical Social Theory*, London: Verso.

Stewart, A. T. Q. (1977) *The Narrow Ground: Aspects of Ulster, 1609–1969*, London: Faber.

Thrift, N. and Forbes, D. (1983) 'A landscape with figures: political geography and human conflict', *Political Geography Quarterly* 2: 247–63.

Yeats, W. B. (1933) *Collected Poems of W. B. Yeats*, London: Macmillan.

9

MAKING SPACE

Gaeltacht policy and the politics of identity

Nuala C. Johnson

INTRODUCTION

The publication of Reg Hindley's book, *The Death of the Irish Language: A Qualified Obituary* (1990), sounded the death-knell for one of the central tenets of cultural policy in Ireland over the last hundred years. Supported by official statistics, census reports and survey material, Hindley suggests that the Irish language no longer has a role in the 'modern' world, that it is doomed to extinction within another generation in the Gaeltacht (Irish-speaking) regions situated along the western coast of Ireland (Figure 9. 1). Contemporary tourist imagery suggests that these same regions are ripe for cultural tourism where 'empty space' – accompanied by narratives of 'empty time' – provide a landscape of consumption for the overseas visitor. The people of these regions are designated as Other – presenting an authentic, primitive escape from modernity for the cultural traveller (O'Connor 1993).

While the west of Ireland has a complex historiography, it has frequently been treated as an homogeneous spatial unit where indices of tradition and modernity can be measured (Johnson 1993). Cultivated through the practice of anthropological, antiquarian and ethnographic research, the western seaboard has witnessed what Agnew (1996: 29) refers to as 'the intellectual genealogy of turning "time into space"'. The conversion of the Gaeltacht regions of Ireland into repositories of a primitive culture, and the attendant cementing of the nation's identity to discourses of premodernism and ethnic purity (Nash 1993), has had important consequences for state policy in the Gaeltacht, and for the response of Gaeltacht people to their assignment as traditional.

This chapter has three main objectives. First, I wish to place the emergence of the Gaeltacht regions as the archive of Irish identity within a broader assessment of nationalist discourse in late nineteenth-century Europe. My purpose is to argue that the drive towards independence, and the relationships articulated between language and cultural identity in Ireland, are exemplars of broader attempts across Europe to establish 'imagined communities' (Anderson 1991). Second, in examining the independent

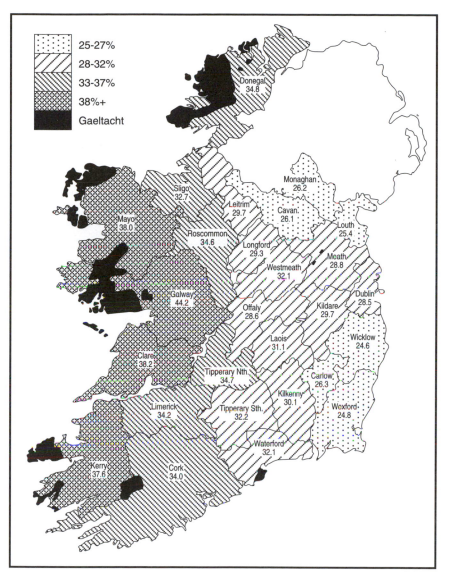

Figure 9. 1 Percentage Irish speakers in 1981 and location of the Gaeltachts
Source: Adapted from Ó Gliasáin 1996

state's policies for language protection, I will suggest that the treatment of Gaeltacht regions as homogeneous places and the direction of economic and regional policy from the political centre precluded any genuine encounter

175

with these linguistic communities as modern and sustainable entities. Finally, I will highlight some of the ways in which the Irish-speaking community in Ireland has sought to escape from the stranglehold of spatial categories, and to develop channels for local democracy that would accommodate and enhance a movement towards cultural equality.

NATIONALISING IDENTITY – LANGUAGE

The role of nationalism in fashioning the European political map has been an important concern of social science over the last thirty years. While there have been numerous attempts to provide definitions, typologies and classifications of nationalism by social scientists (for an overview see Smith 1986), theories of nationalism can be broadly categorised on the basis of those which focus on the cultural features of nation-building originating in the Romantic movement (Kedourie 1966), and those which examine the structural and political antecedents to nationalism and its links with the Enlightenment (Hechter 1979; Breuilly 1982). Both of these approaches suffer from the constraints highlighted by MacLaughlin (1986: 316) where depending on emphasis, 'ethnicity, regional development or uneven development are reified and hypostatized to the extent that they are characterized as major engines of historical change'. Mossé too (1975: 3) has commented that although 'historians have stressed parliamentarianism as being decisive . . . the study of the growth of a new political style connected with nationalism, mass movements and mass politics has been neglected'. Efforts to bridge the gap between studies of nation-building and state-building have centred on identifying the interconnections between the two, particularly at the level of popular participation in politics. Recent analyses of nationalism have consequently begun to incorporate the symbolic as well as the material bases in the evolution of a nationalist politics. In Bhabha's study (1990) of literary 'narrations of the nation', he suggests that there is no single, privileged nationalist narrative. Breuilly (1982: 297) also claims that there is no valid explanatory theory of nationalism, but 'only a number of ways of describing and comparing various forms nationalist politics have taken'.

As consistently acknowledged in this book, Benedict Anderson's account of the origins and the spread of nationalism has been influential, precisely because it emphasises the contingent and 'constructed' nature of national identity in the context of widespread structural change. He states that 'the convergence of capitalism and print technology on the fatal diversity of human language created the possibility of a new form of imagined community, which in its basic morphology set the stage for the modern nation' (1991: 46). The concept of the nation as an 'imagined community' has been furthered through empirical investigation of the emergence of cultural traditions cultivated in the service of the nation. In what Hobsbawm (1983) calls the 'heyday of the invention of traditions' between 1870 and 1914, many of the annual rituals and ceremonies associated with the nation-state were institutionalised and nationalised for the masses. The role of print capitalism, alluded to by Anderson, is supported by Cannadine's analysis of the monarchy in Britain, where 'the massive expansion of the yellow press made it both necessary and possible to present the monarch, in all the splendour of his ritual, in this essentially new way, as a symbol of consensus and continuity to which all might defer' (1983: 133). Geographers too have begun to extend the literature on nationalism, exposing the fluid and contested nature of national identities in a variety of European contexts (Daniels 1993; Gruffudd 1994; Pick 1994).

The saliency of language to issues of national identity came to the fore in nineteenth-century European nation-building. Inspired by the Romantic movement, and particularly by the writings of Johann Gottfried Herder (1744–1803), language came to be seen as an expression of both individual and collective identity, the external badge which would differentiate one 'nation' from another (Kedourie 1966). The politicisation of language in nationalist discourse went hand in hand with state development of national education systems. In France, for instance, the Cahors Committee on Primary Education claimed in 1834 that 'the political and administrative unity of the kingdom urgently required the unity of language in all its parts' (quoted in Weber 1976: 72). Much education policy in nineteenth-century France worked towards that end. Although language was by no means the only source for establishing a collective identity, it did serve as an important cultural signifier in nationalist imaginings for a number of reasons (Fishman 1972). First, language is useful in the cultivation of a national identity because it can make claims to authenticity; it provides a secular symbol which links language communities of the present to those of the past. As this process is conceptualised as a natural phenomenon, language can serve as a legitimation for the 'natural' boundaries of the nation-state. Second, a vernacular literature enables political élites to become centrally involved in a nationalist movement and to make effective connections between themselves and the masses. Finally, Fishman stresses the importance of language planning executed through mass education. He states (1972: 77):

> With the passage of time, and with the control over media and institutions of society, it [language planning] converts the new into the old, the regional into the national, the rural into the urban, the foreign into the indigenous, the peripheral into the central and merely efficient into the authentic.

THE LANGUAGE REVIVAL MOVEMENT IN IRELAND

In the mid-nineteenth century in Ireland, Thomas Davis argued that 'a people without a language is only half a nation' (quoted in Fishman 1972: 49). In this century the central role accorded to language can be seen in the nationalisms of post-colonial states (Williams 1995) as well as in long-established bilingual states (Murphy 1995). Language planning through education has been a cornerstone in debates about language maintenance. By the late eighteenth century, European society witnessed an increased interest in the provision of education. Coupled with changes associated with industrialisation and the democratisation of political life, there was a push towards providing publicly funded education. France, Prussia and Holland were the early innovators in the institution of state-wide national education systems.

In Britain, Ireland served as an early testing ground for current education theory and in 1831 free elementary education was instituted by the Chief Secretary for Ireland, Lord Stanley. While the initial plans for primary education included the intermixing of religious denominations, Coolihan (1981: 2) observes that 'in a climate of hostility and suspicion between the churches and with fears of proselytism rife, Ireland presented a difficult arena for the success of multi-denominational schooling'. The principle of mixed religion schools was quickly abandoned. While the main objectives of primary education were to impart reading, writing and arithmetic skills, Goldstrom's examination of school textbooks (1972) suggests that there was a strong focus on acceptance of the socio-political *status quo*. Coolihan (1981: 21) alleges that 'for much of the century the books contained very little material relating to a distinctively Irish environment and were geared towards the British cultural assimilation policy of the time'. More importantly, perhaps, there was no provision for the use of the Irish language in the national education system in the first forty years of its existence. The decline of the language had, of course, begun earlier than this, but even by 1851 the Census of Population (the first Census published after the devastation of the Great Famine) recorded 300,000 monoglot Irish-speakers and 1.5 million bilingual people (de Fréine 1978). Various language preservation societies were to emerge in the latter half of the nineteenth century, so that by 1879 the Irish language was accepted by the state as one of the optional subjects which could be taught, on payment of a fee, outside school hours.

The real push towards the politicisation of language occurred with the founding of the Gaelic League in 1893. Its precursors included the Society for the Preservation of the Irish Language (1877) and the Gaelic Union (1879), many of the founding members of the League having served their apprenticeships in these societies (Mac Aodha 1972). The objectives of the League resonate with sentiments similar to other European language movements. Indeed, Hutchinson (1987: 117–18) points out that 'these cultural projects were led by secular intellectuals in contact with the international community of Celtic scholars – Rhys in Britain, de Jubainville in Paris and Zimmer in Germany'. Douglas Hyde, Protestant president of the Gaelic League, clearly stated the case:

> The moment Ireland broke with her Gaelic past, she fell away hopelessly from all intellectual and artistic effort. She lost her musical instruments, she lost her music, she lost her games, she lost her language and popular literature and with her language she lost her intellectuality.
>
> (Quoted by Ó hAilín 1972: 96)

The League emerged from a European-wide tradition of antiquarian, scholarly research and its emergence coincided with the other main intellectual movement of the day, the Literary Revival led by W. B. Yeats, Lady Gregory and others. With respect to education policy, the League argued that pupils whose home language was Irish should be taught through the medium of Irish and that English-speaking pupils should be taught Irish within school hours (Ó Túama 1972). Led by an intellectual élite, the League officially professed that it was both a non-sectarian and non-political organisation. It is important to note here that many Ulster Protestants were among its members and by 1899 there were nine branches of the League in Belfast alone (Ó Snodaigh 1995). Garvin (1987: 78) points out, however, that:

> The politicization of culture effected by the League in the early years of the century was to create an official cultural ideology which was arguably hostile to much of the real culture of the community . . . . This official ideology was to dominate much of Irish cultural life for a generation after independence.

The development of an official ideology occurred in part because members of the Gaelic League were later to become leaders in the newly independent state. In particular, Eoin MacNeill – a Northern Catholic – emerged as the first Minister for Education in the Irish Free State. In 1893 he published an article, 'A plea and a plan for the extension of the movement to preserve and spread the Gaelic language in Ireland', in which he articulated what would soon become basic tenets of the League's policy. MacNeill recognised that much of the thrust of the revival movement had emerged from a middle-class intelligentsia and that the revival was being channelled mainly through the formal education system. He suggested that the League must appeal to the mass of the population, and that the movement be organised at the parish level and be localised in focus. This general policy was adopted by the League and while its influence grew slowly at first, it expanded more rapidly once organisers were dispatched around the country

179

to establish branches. A book written by a Gaelic League activist, Fr Eugene O'Growney – *Simple Lessons in Irish* – which focused on both language acquisition and Irish dance and music, was popularised through the branches. Twenty years after its foundation, the League had transformed itself from a small society largely comprising intellectuals to a mass movement with an estimated 1,000 branches and 100,000 members (Redmond-Havard 1913). Ó Fiaich (1972: 63) points out that:

> Perhaps only a few in each branch became fluent speakers of the language, yet during the first two decades of the present century the network of well over a thousand Gaelic League classes which then covered the country was the most highly developed system of adult evening education that Ireland had yet received.

While the leadership of the organisation lay in Dublin, it soon took root countrywide, especially in Munster. As well as being a cultural movement it provided a social outlet for young people in the more rural areas.

In addition to language instruction the League also published magazines and pamphlets, including the *Gaelic Journal* (1904) and *Ireland's Battle for Her Language* (1900) by Edward Martyn. The impact of the League rested in its ability to raise popular consciousness about the Irish language, and to provide one ideological rationale for the Irish independence movement. Ironically, the achievement of independence in some respects deprived the League of its impetus. As Devlin (1972: 87) comments, the greatest blow to the Gaelic League as an organisation 'was the setting of up an independent state, avowedly dedicated to the ideals of the League, as it was the end product of the revolution which the League itself had set in train'. As harbingers of cultural nationalism, the Gaelic League and the Literary Revival both played significant roles in articulating an Irish 'imagined community' and in allowing the West of Ireland to act as a synecdoche of Irish identity. We must be careful to acknowledge, however, that much of the stimulus for this process came from an urban-based environment. Thompson (1967: 66) rightly points out that:

> The Gaelic League was the strongest and most popular [society] for the simple sociological reason that its celebration of folklore appealed to the intellectuals while its moralistic rejection of civilization, decadence and empire appealed to the Catholic lower middle-class.

As the movement towards independence gathered pace, the question of the position of the Irish language within the education system reached the forefront of debate. A bilingual programme in the Irish-speaking districts was introduced in 1904. After the 1918 general election and the establishment of Dáil Éireann in 1919 under Éamon de Valera, a Minister for Irish was appointed as a direct response to the recommendations of the Gaelic League (Ó Búachalla 1988). In August 1921, however, Séan Ó Ceallaigh,

Minister for Irish, was appointed the new Minister for Education and the Irish Ministry was abolished. In the decades preceding the establishment of the Irish Free State the Gaelic League put the Irish language on a solid footing within the education agenda.

DEFINING THE WEST: STATE POLICY IN THE GAELTACHT REGIONS

The real battle for the language, however, has not been in the service of the state, but in what successive governments have recognised as its two principal nurseries – the schools and that dwindling area of the country where Irish is the everyday language of the people.

(Lyons 1971: 627)

From 1922 onwards the Irish Free State had the task of governing an independent state. Part of the new state's strategy was the development of policy in relation to the Irish language. Accompanying this was an emerging image of Irishness, derived primarily from literary sources, which anchored it in the landscape of the western seaboard. As Brown (1981: 83) has commented, in the 1920s 'it was the notion of the virtuous countryman that writers, artists and commentators accepted as the legacy of the Literary Revival period'. This 'virtuous countryman' was cemented onto an image of the Atlantic seaboard where a symbolic unity was imagined to have held sway (Wilson 1977). The fact that the Irish language was strongest in these areas reinforced a sense of cultural harmony.

The state, anxious to fulfil pre-independence cultural policy, responded to the language question by accepting many of the recommendations of the Irish National Teachers Organisation's (INTO) report into the national curriculum for primary education, published in 1922. Central to this report was a criticism of current language policy and a recommendation that the language be taught for one hour per day in all primary schools and that, at infant level, the medium of instruction be in Irish (National Programme Conference 1922). Sufficient teachers, it recommended, should be trained to carry out this task. With MacNeill as new Minister for Education in the Irish Free State, the state readily adopted the INTO's recommendations. McCartney (1973: 86) observes that:

MacNeill's cultural nationalism was almost a copy-book reproduction of that preached by the great European romantic nationalists . . . . Like the European nationalists his appeal was 'to the masses' and Irish should be cultivated 'for the people', 'however poor and struggling', and not for the students.

While the parliamentary debates of the day display an overall commitment to the language, there were a number of dissenting voices – especially with respect to teaching through the medium of Irish and to the precise meaning of 'revival'. While the language was awarded equal status to English in the Irish Free State Constitution, the implementation of a coherent language policy proved difficult, in particular with reference to the recruitment of competent language teachers. Dáil deputies from the Gaeltacht regions were acutely aware of their position as representatives of a cultural reservoir of native speakers. Ó Máille, from Galway, stated in a 1925 Dáil debate that 'I expect any government that will be elected in Ireland will do what it can towards fostering and spreading the Irish language. One of the best ways it can do that is to help higher education in the West' (Dáil Éireann 1925: 1616). Although the Gaeltacht regions had not been officially demarcated on a map, deputies representing these areas were already expressing reservations about the state's commitment to these regions. Williams (1988: 10) alleges that 'language legislation alone would not prove effective in redressing the primarily socio-economic grievances of the "problem region"'. While the language may have been awarded equal legal status, establishing equal cultural status within the new state proved more problematic. Western deputies regularly commented on the lack of infrastructure for their constituents to carry out their daily business with the state through Irish (Johnson 1993).

The first attempt to demarcate the Gaeltacht regions spatially was achieved in the Gaeltacht Commission Report (GCR) which was published in 1926. The Commission was set up by the state's Executive Council with the task of issuing recommendations regarding the use and maintenance of Irish in these regions. Comprised of twelve members – including Dáil deputies, university professors and civil servants – the Commission spent a year preparing its report. It defined an Irish-speaking district as one where over 80 per cent of the population were Irish-speaking, compared to between 25 and 79 per cent in a partly Irish-speaking district. Extrapolating from maps produced by the Congested Districts Board (founded in 1891) and from the 1911 Census of Population returns, the Gaeltacht was officially mapped at the scale of the District Electoral Division.

The net result of this exercise was to fix the Gaeltacht in space and in some respects to define socio-linguistic policy in static geographical categories. In addition to demarcating the Gaeltacht, the Commission made a series of recommendations concerning the protection of the language within these districts. With respect to education provision, the Commission noted the inadequacy of teaching pupils through English and the shortcomings of teaching through Irish by teachers ill-equipped to use the language. Consequently the Commission noted 'that in the child's mind his own language is given a brand of inferiority' (GCR 1926: 11). Thus, while an urban-based intelligentsia may have nurtured an idealistic vision of the West of Ireland, the Gaeltacht inhabitants themselves frequently observed that – both structurally and culturally – they experienced a sense of peripherality.

Public discourses and political practice need not always cohere. To alleviate the problem of a lack of qualified teachers, the Commission suggested the establishment of seven preparatory colleges to be located in the Gaeltacht, with the specific purpose of channelling native speakers into the teaching profession. Although the government did follow this recommendation and drew many of its pupils from Irish-speaking areas, the colleges themselves were seldom located in the Gaeltacht (Johnson 1992). With respect to the use of Irish in public administration, the Commission noted that:

> Detailed instructions as to its use in administration, either in Irish Speaking Districts or elsewhere, have not been issued. No department is charged with the duty of seeing that the National language is given any preference in use to English, even in the Gaeltacht or that it is used at all.
>
> (GCR 1926: 25)

No separate government department with responsibility for this area of state cultural policy was established until 1956.

It is, however, with respect to the economic conditions prevalent in the Gaeltacht that the Commission reserved most comment. The Commission did not treat the Gaeltacht as a single, homogeneous unit but specifically highlighted the existence of 'special areas' where the economic conditions were even more serious than in the region as a whole. The Report stated that:

> The problem which exists in these areas is, from the point of view of congestion, very serious – so serious in fact that, hitherto, every responsible authority has hesitated to approach it . . . . These populations have been almost entirely excluded, in the past, from the operation of economic land settlement and migration.
>
> (GCR 1926: 45)

The Commission recommended the development of local, indigenous industries such as fishing and textiles, the breaking up of grasslands and their resettlement solely by Irish-speaking families from the more congested districts. These were radical proposals that clearly made connections between cultural policy and regional economic policy but the government rejected most of them (Johnson 1993). The Commission's suggestion that a special commission be established to oversee and evaluate government policy in relation to the Gaeltacht was dropped. Instead, a Minister for State 'by a scheme of close co-operation with the various Departments responsible for education, agricultural instruction, housing, health etc. will provide the special attention and co-ordination which is necessary' (*Statement of Government Policy on Recommendations of the Commission* 1928: 29).

Thus while the Commission completed the task of putting a boundary around the Gaeltacht, the instruments proposed to maintain or expand that boundary were largely rejected by the state. Having defined the problem as a geographical one, the state was reluctant to provide a geographically based solution (Johnson 1993). National policy took precedence over regional policy in the state's handling of the Gaeltacht until the 1950s. The maps that the Gaeltacht Commission produced served, ironically, as the blueprint for monitoring the contraction of the Gaeltacht. An analysis of the public discourse about the West of Ireland and the Gaeltacht in particular, acting as repositories of Irish identity in the twentieth century, needs to be matched by a critical examination of actual state policy during this time. The language of nationalism may not always be reflected in the practice of government.

RESPONSES FROM THE GAELTACHT

State aid to the Gaeltacht districts up to the 1950s focused on agricultural improvement, the development of indigenous industries, housing improvement and some support for educational provision. Nevertheless, although the Gaeltacht had been defined as a regional problem area, there was no separate state apparatus designated to deal with it, nor was there much by way of consultation at local level. In 1956, however, a separate Ministry was established in Dublin, with responsibility for administering government policy with respect to the Gaeltacht (Hindley 1990). In addition the state, influenced by the economic theory of the day regarding peripheral areas, established a semi-state body, Gaeltarra Éireann, to promote the industrialisation of the Gaeltacht regions. The primary role of Gaeltarra was the provision of employment in the industrial sector. Initially focusing on developing indigenous industries (for example, knitting), in 1965 Gaeltarra's powers were extended to give grant-aid to new industries and to enter partnerships, through share capital, with private enterprises.

Particularly targeting overseas companies, the agency enjoyed moderate success in the establishment of small industrial estates in the Gaeltacht. Numbers employed in enterprises supported by Gaeltarra increased from 700 in the 1960s to 4,300 in the 1980s (Commins 1988). Gaeltarra's strategy, however, presented a number of problems. First, the management teams in many of these new industrial enterprises were directly recruited from outside the Gaeltacht and, consequently, were more likely to be English-speakers. As one Gaeltacht activist commented:

> The importing of foreign capital and expertise in management, marketing and skills to the Gaeltacht, which Gaeltarra Éireann has achieved, should be welcomed just now, but if it continues to be the main form of Gaeltacht economic development there is a danger that it contains within itself the seeds of eventual Gaeltacht cultural and linguistic decay.
>
> (Quoted in Johnson 1979: 70)

Second, the management of these branch plants from overseas headquarters led to the absence of local input in discussions about their economic future. Third, the centralised nature of the statutory powers and government agencies invested with these powers divorced the overall planning of the Gaeltacht from the people living there. Finally, and most significantly from the point of the communities, Gaeltarra Éireann did not have an active cultural or linguistic policy. Its role as a state-funded regional policy organisation for a specified geographical area was not matched by a clear set of cultural objectives. By the 1960s the Gaeltacht had, to all intents and purposes, been redefined as a problem region, peripheral and backward, rather than as the cultural heartland of the 'nation'.

One of the responses of Gaeltacht people to these processes was the establishment of the Gaeltacht Civil Rights Movement (Cearta Sibhialta na Gaeltachta) in the late 1960s. A radical shift emerged from within the Gaeltacht itself and attention was focused on issues related to civil rights and citizenship within the Irish state. In common with civil rights movements in North America and Europe, the terms of the public debate in part shifted away from narrow debates about nationhood to broader questions related to equal opportunity and civil liberties. The movement

> advocated the view that economic development could be more deliberately based on the development of natural resources, that it could be structured to allow for greater public participation, and moreover, that it could be better harmonized with the social and cultural circumstances of the Gaeltacht.
>
> (Commins 1988: 17)

From the point of view of 'imagined communities', the Gaeltacht Civil Rights Movement challenged the state's right to dictate policy without due consultation with the people subjected to the policy.

The development of a locally based co-operative movement was one principal consequence of this challenge. While the co-operative movement had emerged in the late nineteenth century and had been adopted with some success in the dairying regions of the country, it had little impact on the Gaeltacht (Bolger 1977). When in 1967 the first two co-operatives were established in west Kerry and west Mayo, the government reacted by awarding them formal recognition and by establishing a Gaeltacht co-operative scheme whereby the state supported local initiative through the provision of start-up grants. What differentiated the co-operatives from other aspects of Gaeltacht policy was that the impetus for their establishment was local. As Johnson (1979: 71) points out, the most important feature of the movement 'is that momentum comes from within the Gaeltacht itself'. By 1979, there were twenty-two Gaeltacht co-operatives. While their success varied from place to place, the co-operatives represented a move closer to bottom-up economic and cultural policy, a shift from the

185

top-down policies which had preceded them since the foundation of the state (Duffy and Breathnach 1983). Unlike the co-operatives in other parts of the country, those in the Gaeltacht were not just agriculture-based but also included the provision of local infrastructure, the development of tourism potential and resource management. In their survey of community perspectives on development in the Kerry and Galway Gaeltachts, however, Duffy and Breathnach (1983: 63) warn that 'despite the general goodwill displayed towards them, they have largely failed to transcend their popular image as yet another agency – albeit locally based – which "delivers" development to a client community'. This view may be due to the lack of familiarity with co-operative principles among the Gaeltacht population. Johnson (1979) observed that the vast majority of the shareholders in the co-operatives which she analysed were those directly benefiting from their activities. In addition, Commins (1988: 18) points out that 'statutory agencies cannot easily accommodate or aid non-statutory structures which engage in "defiant" development . . . [and] challenge the implicit assumptions of socio-economic policies'.

Another major change brought about by the activities of the Gaeltacht Civil Rights Movement was in the constitution and remit of state agencies dealing with the Gaeltacht. In 1980 Gaeltarra Éireann was replaced by Údarás na Gaeltachta (the Gaeltacht Authority). Its headquarters were located in Cois Fharraige (in the Galway Gaeltacht). This new agency differed from its predecessor insofar as it gave democratic representation to the people of the Gaeltacht. Building on the arguments made by Ceart Sibhialta na Gaeltachta, the Board of Údarás na Gaeltachta was comprised of seven members (from a total of thirteen) directly elected by Gaeltacht voters. It was hoped that this strategy would empower Gaeltacht people and give them greater representation in the planning process for their own area. Údarás na Gaeltachta was also to have a more active policy towards language maintenance. Its role was defined as follows:

> An tÚdarás shall encourage the preservation and extension of the Irish language as the principal medium of communication in the Gaeltacht and shall ensure that Irish is used to the greatest extent possible in the performance by it and on behalf of its functions.
>
> (Cited in Commins 1988: 16)

In terms of industrial development Údarás continued the brief of Gaeltarra, and there was some extension of powers in terms of promoting the Gaeltacht for investment. Central government, however, has continued to maintain some control. Grants above £500,000 must receive approval by the Minister for the Gaeltacht. Although Údarás na Gaeltachta's responsibility was extended to include linguistic, cultural, social, physical and economic objectives, ultimately the major role of the Údarás is job creation through industrial development. In a survey of people in the Kerry and Galway

Gaeltachts, Duffy and Breathnach (1983) noted that respondents saw the Gaeltacht Authority and the co-operatives as alternative means to the same end – namely, the provision of employment in the Gaeltacht. Most notably, 'very few people appear to be actively aware of, or concerned about, the potential function of either organisation as an agency for linguistic/cultural development' (Duffy and Breathnach 1983: 62). Criticisms of Údarás na Gaeltachta focus on the fact that it never became the all-embracing local authority originally envisaged by those who lobbied for it (Ó hÉallaithe 1989) and perhaps this fact accounts for the public's perception of its role. The economic performance of industries supported by Údarás have also experienced mixed success (*Irish Business* 1981). Ó Cinnéide *et al.* (1985), in their analysis of the effects of industrialisation on language shift, have observed that industrialisation in itself need not have an adverse effect on the language provided that labour force recruitment positively discriminates in favour of Irish-speakers. However, when outsiders are recruited to the Gaeltacht, they 'should be at least sympathetic to the Irish culture' (Ó Cinnéide *et al.* 1985: 14). Údarás has, in principle, tried to apply these criteria in its job-creating exercises and it has supported the establishment of Irish-language pre-school play groups and youth clubs. Despite its activities, this has not stopped the persistent out-migration among young adults, but this pattern may partly help account for the significant expansion in the 1980s of Irish-language schools in urban centres such as Dublin, Cork, Limerick and Belfast.

Finally, the Gaeltacht Civil Rights Movement was also instrumental in bringing about the establishment of Raidió na Gaeltachta, a radio station dedicated to broadcasting to the Irish-speaking community nation-wide. The radio station has had the effect of reducing the importance of the geographical separation of the Gaeltachts by connecting areas cut off from each other spatially, and thus creating a network where Irish speakers can exchange news and views that extend beyond the strictly local. The imminent introduction of Teilifí's na Gaeilge represents another instance of the effectiveness of pressure-group politics from within the Gaeltacht and among Irish speakers outside the Gaeltacht.

THE DEATH OF THE IRISH LANGUAGE REVISITED

For the last one hundred and fifty years the death of the Irish language has been forecast. The most recent testament to this is Hindley's (1990) analysis of language decline from the nineteenth century onwards. Relying primarily on published official statistics and secondary sources, Hindley's discussion focuses on the inevitable decline of the language, given the historical economic processes necessitating a conversion to English. His empirical focus is centred on the Gaeltacht regions themselves as they have been defined by the state over the last seven decades. This reliance on geographical definitions of language usage overlooks the use of Irish outside the Gaeltacht, ignores the dynamic processes involved in language shift and underestimates the complex spatial interconnections between the Gaeltacht and other places in the late twentieth century. (However, maps displaying proportions of Irish speakers based on nineteenth- and twentieth-century census data are at best approximations and at worst crude over-simplifications.) Moreover, as Ó Ciosáin (1991: 7) tellingly points out, 'the Gaeltacht is not a place, or is not only a place, it is the people, the community of native Irish speakers'. That community exists inside and outside the fixed boundaries of the map. While mapping has long preoccupied the state in its articulation of cultural policy, for the state the map has frequently acted as an archive to monitor decline and failure. The map then acts as a metaphor for a failed cultural project, a spatially defined society imploding under the strains of modernity. In this sense the map encloses as much as it discloses. A new map of the Gaeltacht, however, would extend beyond the boundaries of the West of Ireland or indeed of the state itself to include diaspora in Britain, the United States, Australia and beyond. Stuart Hall (1995: 48) suggests that diaspora are 'people who belong to more than one world; speak more than one language (literally and metaphorically); inhabit more than one identity; have more than one home'. In the late twentieth century, Hall's observations may have more relevance for those living inside and outside the Gaeltacht than any notion of them being repositories of 2,500 years of civilisation.

Recent writing on political identity has suggested that it too is not fixed as forcefully in time and space as older discourses have proposed (Jackson 1989). In the opening sections of this chapter I suggested that the West of Ireland was 'invented', primarily by an intelligentsia, as a spatial metaphor for Irish nationhood. Bonded by language, landscape, economic circumstance and a history of marginalisation (both before and after independence), the lands along the western seaboard were homogenised intellectually along an axis of tradition and modernity. In common with other European attempts at nation-building, an idealised landscape and people were invoked to represent the essence of the nation. The fact that people in western Ireland spoke Irish reinforced its status as a metaphor for cultural identity. Ironically, the practices of successive governments since independence have not always been in tune with the public rhetoric about cultural identity.

The revisionist theme in Irish historiography has begun to raise some theoretical and methodological questions regarding the processes of recounting the Irish past and, in particular, nationalist interpretations of it (Brady 1994). Although revisionist writing has had comparatively little to say about the role of the Irish language in identity formation or in contemporary political debate, the upsurge in local studies that has accompanied revisionism (see Chapter 1) offers new avenues of research that move beyond state-centred scales of analysis. Indeed, Whelan (1992) highlights the usefulness of locally based research in unravelling the complex relationships between official and popular histories. In the context of the Irish language, Lee (1989) stresses the relevance of comparative research on language usage and bilingualism. Focusing on European states of similar size to Ireland, he challenges the economic arguments that have frequently been used to support the exclusive use of English in Ireland. Recent research also highlights the role of the Irish language in the Presbyterian tradition (Blaney 1996). Taken together, in-depth local studies and a comparative perspective may shed greater light on the recent increase in Irish language usage, outside the Gaeltacht and in Northern Ireland, than more traditional methodologies. By adopting some of the new perspectives in cultural geography which advance multicultural and multivocal approaches, discussions of the Irish language can move beyond interpretations of language which connect it solely to a nationalist politics. Questions dealing with the relationship between language and class, gender or ethnicity might be fruitfully explored so that we, as academics, may have a fuller understanding of those that we seek to represent.

ACKNOWLEDGEMENT

Many thanks to Máirín Nic Eoin for her helpful comments on an earlier draft of this chapter.

REFERENCES

Agnew, J. (1996) 'Time into space: the myth of "backward Italy" in modern Europe', *Time and Society* 5, 1: 27–45.

Anderson, B. (1991) *Imagined Communities: Reflections on the Origin and Spread of Nationalism*, revised ed., London: Verso.

Bhabha, H. (1990) (ed.) *Nation and Narration*, London: Routledge.

Blaney, R. (1996) *The Presbyterians and the Irish Language*, Belfast: Ulster Historical Foundation.

Bolger, P. (1977) *The Irish Co-operative Movement, its History and Development*, Dublin: Institute of Public Administration.

Brady, C. (1994) (ed.) *Interpreting Irish History: The Debate on Historical Revisionism*, Dublin: Irish Academic Press.

Breuilly, J. (1982) *Nationalism and the State*, New York: St Martin's Press.

Brown, T. (1981) *Ireland: A Social and Cultural History, 1922–79*, London: Fontana.

Cannadine, D. (1983) 'The context, performance and meaning of ritual: the British monarchy and the "invention of tradition", *c.* 1820–1977', in E. Hobsbawm and T. Ranger (eds) *The Invention of Tradition*, Cambridge: Cambridge University Press.

Commins, P. (1988) 'Socioeconomic development and language maintenance in the Gaeltacht', *International Journal of Sociology of Language* 70: 11–28.

Coolihan, J. (1981) *Irish Education: History and Structure*, Dublin: Institute of Public Administration.

Daniels, S. (1993) *Fields of Vision: Landscape Imagery and National Identity in England and the United States*, Cambridge: Polity.

Dáil Éireann (1925) *Official Report*, 8, Dublin: Stationery Office.

de Fréine, S. (1978) *The Great Silence*, Cork: Mercier.

Devlin, B. (1972) 'The Gaelic League – a spent force?', in S. Ó Túama (ed.) *The Gaelic League Idea*, Cork: Mercier.

Duffy, P. J. and Breathnach, P. (1983) 'Community perspectives on Gaeltacht development: a review of some research findings', in P. Breathnach (ed.) *Rural Development in the West of Ireland: Observations from the Gaeltacht Experience*, Occasional Paper 3, Maynooth College: Department of Geography.

Fishman, J. (1972) *Language and Nationalism*, Rowley, MA: Newbury House.

Gaeltacht Commission (1926) *Report*, Dublin: Stationery Office.

Garvin, T. (1987) *Nationalist Revolutionaries in Ireland 1858–1928*, Oxford: Clarendon.

Goldstrom, J. M. (1972) *The Social Content of Education, 1808–70: A Study of Irish School Textbooks*, Shannon: Irish University Press.

Gruffudd, P. (1994) 'Back to the land: historiography, rurality and the nation in interwar Wales', *Transactions, Institute of British Geographers* NS 19: 61–78.

Hall, S. (1995) 'New cultures for old', in D. Massey and P. Jess (eds) *A Place in the World: Places, Culture and Globalization*, Oxford: Oxford University Press.

Hechter, M. (1979) *Internal Colonialism: The Celtic Fringe of British National Development, 1536–1966*, California: University of California Press.

Hindley, R. (1990) *The Death of the Irish Language: A Qualified Obituary*, London: Routledge.

Hobsbawm, E. (1983) 'Introduction: inventing traditions', in E. Hobsbawm and T. Ranger (eds) *The Invention of Tradition*, Cambridge: Cambridge University Press.

Hutchinson, J. (1987) *The Dynamics of Cultural Nationalism: The Gaelic Revival and the Creation of the Irish Nation State*, London: Allen and Unwin.

Irish Business (1981) 'Údarás na Gaeltachta – attracting lame ducks', 2–9 March.

Jackson, P. (1989) *Maps of Meaning*, London: Routledge.

Johnson, M. (1979) 'The co-operative movement in the Gaeltacht', *Irish Geography* 2: 68–81.

Johnson, N. C. (1992) 'Nation-building, language and education: the geography of teacher recruitment in Ireland, 1925–55', *Political Geography Quarterly* 11, 2: 170–89.

—— (1993) 'Building a nation: an examination of the Irish Gaeltacht Commission Report of 1926', *Journal of Historical Geography* 19, 2: 157–68.

Kedourie, E. (1966) *Nationalism*, 3rd ed., London: Hutchinson.

Lee, J. J. (1989) *Ireland, 1912–1985: Politics and Society*, Cambridge: Cambridge University Press.

Lyons, F. S. L. (1971) *Ireland Since the Famine: 1850 to the Present*, London: Weidenfeld and Nicolson.

Mac Aodha, B. (1972) 'Was this a social revolution?', in S. Ó Túama (ed.) *The Gaelic League Idea*, Cork: Mercier.

MacLaughlin, J. (1986) 'The political geography of nation-building and nationalism in the social sciences: structural vs. dialectical accounts', *Political Geography Quarterly* 5, 4: 299–329.

McCartney, D. (1973) 'MacNeill and Irish-Ireland', in F. X. Martin and F. J. Byrne (eds) *The Scholar Revolutionary: Eoin MacNeill 1867–1945 and the Making of the New Ireland*, New York: Harper and Row.

Mossé, G. (1975) *The Nationalization of the Masses*, New York: Howard Fertig.

Murphy, A. (1995) 'Belgium's regional divergence: along the road to federation', in G. Smith (ed.) *Federalism: the Multi-ethnic Challenge*, London: Longman.

Nash, C. (1993) ' "Embodying the nation": the west of Ireland landscape and Irish identity', in B. O'Connor and M. Cronin (eds) *Tourism in Ireland: A Critical Analysis*, Cork: Cork University Press.

National Programme Conference (1922) *National Programme of Primary Instruction*, Dublin: Brown and Nolan.

Ó hAilín, T. (1972) 'Irish revival movements', in S. Ó Túama (ed.) *The Gaelic League Idea*, Cork: Mercier.

Ó Búachalla, S. (1988) *Education Policy in Twentieth-Century Ireland*, Dublin: Wolfhound Press.

Ó Cinnéide, M., Keane, M. and Cawley, M. (1985) 'Industrialization and linguistic change among Gaelic-speaking communities in the West of Ireland', *Language Problems and Language Planning* 9, 1: 3–15.

Ó Ciosáin, E. (1991) *Buried Alive: A Reply to the Death of the Irish Language*, Baile Átha Cliath: Dáil Uí Chadhain.

O'Connor, B. (1993) 'Myths and mirrors: tourist images and national identity', in B. O'Connor and M. Cronin (eds) *Tourism in Ireland: A Critical Analysis*, Cork: Cork University Press.

Ó hÉallaithe, D. (1989) 'Gaeltacht experiment in democracy falters', *Alpha*, 11 May.

Ó Fiaich, T. (1972) 'The great controversy', in S. Ó Túama (ed.) *The Gaelic League Idea*, Cork: Mercier.

Ó Gliasáin, M. (1996) *The Language Question in the Census of Population*, Dublin: Institiúio Teangeolaíochta Éireann.

Ó Snodaigh (1995) *Hidden Ulster: Protestants and the Irish Language*, Belfast: Lagan Press.

Ó Túama, S. (1972) (ed.) *The Gaelic League Idea*, Cork: Mercier.

Pick, D. (1994) '*Pro patria*: blocking the tunnel', *Ecumene* 1: 77–94.

Redmond-Havard, L. (1913) *The New Birth of Ireland*, London: Collins.

Smith, A. D. (1986) *The Ethnic Origin of Nations*, Oxford: Blackwell.

Statement of Government Policy on Recommendations of the Gaeltacht Commission (1928) Dublin: Stationery Office.

Thompson, W. (1967) *The Imagination of an Insurrection: Dublin Easter 1916: A Study of an Ideological Movement*, Oxford: Oxford University Press.

Weber, E. (1976) *Peasants into Frenchmen: the Modernization of Rural France 1870–1914*, Stanford: Stanford University Press.

Whelan, K. (1992) 'The power of place', *Irish Review* 12: 13–20.

Williams, C. (1988) (ed.) *Language in Geographic Context*, Philadelphia: Multilingual Matters.

Williams, C. (1995) 'A requiem for Canada?', in G. Smith (ed.) *Federalism: The Multi-ethnic Challenge*, London: Longman.

Wilson, J. (1977) 'Certain set apart: the Western Island in the Irish Renaissance', *Studies* 66: 264–73.

10

THE IMAGINING OF PLACE

Representation and identity in contemporary Ireland

Brian Graham

INTRODUCTION

In this chapter, I extend the discussion of identity to examine the contrasting roles played by the contested imagery of cultural landscape in defining or impeding social cohesion in Ireland. As explained in Chapter 1, manipulated depictions of landscape offer an ordered, simplified vision of the world and act as a system of signification supporting the authority of an ideology and emphasising its holistic character. These constructs are central to discourses of inclusion and exclusion and to definitions of the Other and Otherness. The ubiquitous relationship between politico-cultural institutions and territoriality suggests that agreed representations of place are fundamental to establishing the legitimacy of contemporary authority which is derived, not from the support of a numerical majority alone, but through renditions of plurality that transcend class, gender and ethnic divisions.

In addressing the particular significance of emblematic place to understanding the contested nature and meanings of identity in contemporary Ireland, this discussion incorporates many of the themes already examined in the previous chapter, albeit largely from the perspective of Northern Ireland. Although unionism is very much a fractured concept, its adherents remain largely defined by a shared negativity. This is expressed in adversarial Otherness to the Republic of Ireland, to which Ulster Protestants react with a sense of inferiority and defensiveness, mostly stemming directly from ignorance of the Irish past, combined with a sense that history is being used against them in claims to the moral high ground (Pollak 1993: 97). The mainstream unionist version of Britishness is equally flawed, particularly in its failure to recognise the conditional and contested nature of the British state.

Gallagher (1995), who argues that there are three nations in Ireland – an Irish nation, an Ulster Protestant nation and part of the British nation – assumes that the Irish nation can be defined sufficiently broadly to encompass both Ulster nationalists and the population of the Republic of Ireland. It is argued here, however, that the contemporary renegotiation of Irish identities has widened the dichotomy between North and South, irrespective of ethnic alignment. Both unionist and nationalist identities in Ulster remain heavily informed by representations of nineteenth-century ethnic nationalism, later incorporated into the 1937 Constitution as the moral core of the Irish state (Lee 1989: 648). This required a representation of place, which denied heterogeneity in the interests of a communal solidarity that subsumed the ethnic plurality present in the South before independence. Economically disastrous in the long term, this partisan ideology still provided the fledgling independent state with a strength and symbolic unity of purpose that contrasts markedly with the unagreed nature of Northern Ireland. Today, however, this hegemonic representation is increasingly irrele-vant as the Republic is transformed into an energetic, outward-oriented member of the European Union and a markedly more secular state, in which the exclusivity of ethnic nationalism is gradually being replaced by the inclusiveness of civic nationalism with its notions of a people linked by a communality of laws and institutions of citizenship rather then sectarian ethnic markers (see Chapter 7).

In summary, I argue here that the absence of an agreed representation of place, congruent with territory, to which its inhabitants can subscribe irre-spective of their class, gender or ethnicity, is a primary factor distinguishing North and South in contemporary Ireland. The result is particularly evident in the troubled nature of unionist and nationalist identities in Ulster. I am concerned too with the ways in which representations of landscape and place create manipulated geographies that mesh landscape and memory within the contested arenas of cultural identity and nation-building. Inevitably, these landscape texts are concerned with mutual discourses of inclusion and exclu-sion, based on antagonism to the Other. They are constructed to act as signifiers of particular discourses within the welter of contested identities that is modern Ireland.

LANDSCAPE AND MEMORY

The embodiment of public memory in landscape provides a robust example of the ways in which representations of place are intimately related to the creation and reinforcement of official constructions of identity and power and to the whole question of empowerment. These mythical worlds become literal (Agnew 1996: 35), even though they may bear little relationship to the places in which most people who subscribe to the mythology actually live. Memory can be outer- or inner-directed but, whichever, it too is a social construct, in this context a direct parallel to the dual linear narratives of history that were imposed on a multifarious Irish identity in the late nine-teenth century (see pp. 57–60). Samuel (1995) regards memory, not as timeless tradition, but as being transformed from generation to generation through, for example, the contrived nature of heritage, which can be defined, not as artefacts and traditions inherited from the past, but by the contested modern meanings that are attached to these objects (Tunbridge and Ashworth 1996). The function of memory is defined by the present, its connections with history and place vested in emblematic landscapes and places of meaning that encapsulate public history and official symbolism.

One illustration is provided by the varying attitudes in Ireland to landscapes of remembrance. These too are complex social constructions that can be read in a variety of ways. For instance, Heffernan (1995) argues that the war memorials and cemeteries of the Western Front – muted, serene, peaceful and intensely moving – convey no real sense of sacrifice to the nation-state. Instead, they are immortal, sacred landscapes, essentially apolitical. Unionists, however, regard them as symbolic of Ulster's embattled past, thereby fulfilling some part of the need for an outer-directed memory. The slaughter of the 36th (Ulster) Division on 1 July 1916, the opening day of the Battle of the Somme, is central to unionist mythology as the debt that Britain owes. Thus while the Somme Heritage Centre, opened in 1994 near Newtownards, County Down, is predicated 'upon the moral necessity of remembering the dead', there is a clear tension between this role and the simultaneous renditions of the events which it records in competing political discourses. In the unionist state, Ulster's sacrifice for Britain became the leitmotif of loyalty (Officer 1995) while, in the South and nationalist North, public remembrance of Irish deaths in the Great War became little more than a peripheral embarrassment (Leonard 1996). Remembrance Day is seen as a unionist ceremony. The carnage of the Somme and the other battlegrounds of the Western Front can also be read, however, as a memory of shared loss, the sacrifice of the 'sons of Ulster' matched, for example, by that of the mainly Catholic 16th (Irish) Division around Messines in the several Battles of Ypres. The Ulster Tower at Thiepval, the memorial to the 36th Division, can thus become an inner-directed mnemonic symbol of the mutual suffering of Protestants and Catholics (Graham 1994a).

This example demonstrates how continuously renegotiated landscapes of memory are implicated in the construction and maintenance of cultural identities. Johnson (1995) distinguishes between the construction of landscape images through the imaginings of an intellectual élite of writers, politicians, artists and architects, and 'popular' imagined communities. The essence of an élitist narrative of place is encapsulated in Ashworth's argument (1993) that dominant ideologies create specific place identities, which reinforce support for particular state structures and related political ideologies. Although these constructs are transmitted in many ways – notably through political structures, education, socialisation and media – the representative landscape is also a substantial device in the evocation of official collective memory. Johnson (1994) points to the importance of monuments and statuary in Ireland as one important means of arousing public imagination. However, it must be recognised that these monuments, like all heritage artefacts, are polyvocal – capable of expressing a multiplicity of political ideologies (Barnes and Duncan 1992) – and thus symbolic of the multifaceted nature of Irish political identity.

Nevertheless, one vision may acquire hegemonic status during a particular epoch as 'time [is] translated into space . . . "blocks" of space [being] labelled with the essential attributes of different time periods relative to the idealized historical experience of one of the blocks' (Agnew 1996: 27). Consequently hegemony, which can best be visualised as a dominant cultural form accepted as legitimate in that it embodies the aspirations of a society, is an active process that is constantly re-articulated and renegotiated as historical circumstances alter. Duffy (1994) demonstrates, for instance, how one collective hegemonic memory of eighteenth-century Ireland was shaped by the valued or preferred landscapes of the landowning upper echelons of society and wealthy tourists. Although many writers commented on the poverty and squalor of rural Ireland's teeming population, visual art stressed the wildness and beauty of the 'natural' landscape, or the demesnes and big houses of the members of the landowning class, which commissioned the work and whose values and tastes it incorporates.

In contrast, the nineteenth-century Gaelic Revival – initially, if ironically, also associated with the Anglo-Irish élite – created an emblematic landscape in which certain artefacts acquired mnemonic status because they fulfilled the need for a retroactive continuity of culture to a distant age prior to the 'book of invasions' that Anglophobic Irish history all too often became. As shown in Chapter 4, the imagery accompanying this narrative was of a predominantly rural Ireland, its true cultural heartland defined by the landscapes and way of life of the wild, western Atlantic fringes – those furthermost removed from Anglo-influences but also most congruent with the precepts of nineteenth-century Romanticism. Iconic sites of continuity in this mythology included Celtic monasteries, Iron Age hill-forts and megalithic tombs. There was no place for towns, archetypal symbols of the Other and dismissed as an alien (and particularly English) innovation. This narrative of place – like the nationalist rhetoric from which it emanated – became that of 'one nation' Irish-Ireland, its ultimate corollary an exclusion of any social groups not encompassed within an ideology eventually wholly Gaelic and Catholic in ethos.

Compare the renegotiation of this official and élitist landscape of memory to present-day Belfast. Here popular 'imagined communities' and life-or-death landscapes of fear are marked and reinforced by flags, murals, painted kerbstones and graffiti, and claimed by marching (Rolston 1991). These cultural signifiers embody memory – wall murals, for example, often entrench existing structures and beliefs rather than advocating any potential transformation toward a new Ulster (Jarman 1992; Bryson and McCartney 1994). Freedom Corner in Protestant east Belfast and Free

Derry Corner in the city's Catholic Bogside do not seek consensus but merely echo the mutual incomprehension of the question: whose freedom? To Edna Longley (1991: 37), such symbols are an inner-directed mnemonic, a rhetoric of memory that tries to place the past beyond argument. Again, marching can be depicted in similar terms as a territorial marker, justified by its connections to the historical events being commemorated, the very routes an expression of communal consciousness and solidarity. Orangemen would prefer to ask nobody's consent to their marches. However, loyalist parades seem to be increasing in number (Jarman and Bryan 1996), while 'traditional' routes are often more flexible and the history of many marches less continuous than is alleged by loyalists proclaiming that their cultural traditions are under threat.

What all this emphasises is that the same components of human landscape acquire contrasting meanings as hegemony is negotiated or contested in both official and popular discourses. Like their predecessors, current myths also evoke and sanctify memories of the past. It is we who impose our narratives on that past and it is we who construct it into our collective and individual memories, which are then played out through our manipulated constructions of place. These texts are embodied in our emblematic landscapes, whether official or unofficial. The mural on a Belfast gable, the official state monument, the heritage artefact, the patriot's grave and the war cemetery are all parts of landscapes of memory, legitimating our presents by connecting them to the conflicting justifications of our pasts. That the same places may participate in different landscapes, denoted by different meanings, merely reflects the unagreed nature of our society.

THE DIVERSE UNITY OF IRELAND

The notion that landscapes embody memory in discourses of inclusion and exclusion is closely linked to the idea that manipulated geographies also function as symbols of identity, validation and legitimation. Thus there are archetypal national landscapes, which draw heavily on geographical imagery, memory and myth (Gruffudd 1995). Continuously being transformed, these encapsulate distinct home places, defined by their very difference to the Other. The 'imagined community' comprises people who are bound by cultural and, more explicitly, political networks, all set within a territorial framework that is defined through whichever traditions are currently acceptable, as much as by its geographical boundary. As we have seen, those national traditions are narratives that are invented and imposed on space.

One of the critical contemporary distinctions distinguishing South from North in Ireland concerns the contrasting official symbolic universes created through narratives of place identity. The fledgling Irish Free State derived strength, legitimacy and a unity of purpose from its exploitation of the hegemonic imagery of the West of Ireland as Ireland's cultural heartland.

Nevertheless, the constitutional institutionalisation of de Valera's ideal of an agrarian, homely, Catholic society could never accommodate Protestant, industrialised north-east Ireland. To Irish-Ireland, it became the lost land, shrouded in Celtic mists and populated with warrior heroes, the most intensely Irish of all the regions of the island. The success of the Roman Catholic church in Ireland in harnessing itself to this version of history also ensured that the decisive ethos within nationalist Ireland after 1922 was Catholic rather than Gaelic (Ó Tuathaigh 1991: 63). Faced with this ideological victory in which everything Irish was sequestered as Republican and Catholic, Protestants – even those opposed to unionism – increasingly lost, abandoned, or were excluded from any sense of being Irish. Moreover, the unionist state in the North failed to develop an alternative indigenous cultural synthesis, relying instead on the political dimensions of the union to delimit Northern identity. Undermined by the ambiguities of that relationship, the poverty of unionist historical awareness and a political unwillingness to develop a representation of Northern Ireland that transcended the sectarian dichotomy, the result has been cultural incoherence and political impotence (Brown 1991: 82).

As we have seen, traditional Irish nationalism involved the deification of places essentially defined by Daniel Corkery's realisation of a 'Hidden Ireland', in which Irish identity was couched in terms of a Gaelic society of great antiquity oppressed by British economic, political and religious discrimination (Cullen 1988). MacLaughlin (1993) argues that the dominance of this imagery reflected the political and moral hegemony in Irish society of the petty bourgeoisie and other like-minded social groups – substantial farmers, local business interests and the Roman Catholic church. It was their interests and values that came to define a new Irish state which had little to offer Protestants. Therefore the rendition of Ireland, enshrined in the 1937 Constitution, was but one particular socially constructed trope of exclusivity which, having outlived its epoch of genesis, is no longer an appropriate expression of collective Irish memory. Hegemonic ideas are being renegotiated and refashioned in the multifaceted context of secularisation, Europeanness and the seemingly eternal conflict in the North. Nevertheless, the traditional rendition of identity is perpetuated by political conservatism, tourism imagery and the folk memories of the diaspora. Crucially, it still continues to inform both Ulster unionist and nationalist representations of Irishness.

If historians and cultural theorists remain enmeshed in the controversies of revisionism and post-colonialism, geographers – as argued in Chapter 1 – have often seemed intellectually more inclined to accept what has become the essentially post-colonial representation of a geographically heterogeneous Ireland, the personality of which is largely defined by its multiplicity of regional differences – among which Ulster's particularity is but one (Graham and Proudfoot 1993). Denying the exclusivity of post-partition Irish place and the emphasis on continuity within the Gaelic rhetoric, this rendition demands the dismantling of Ireland into the narrative of regional variety which Smyth has outlined in Chapter 2. Thus Whelan (1992, 1993) questions the whole myth of homogeneity, arguing that Ireland was and remains an island comprised of localised regions, an interpretation which demands the deconstruction of the potentially divisive nature of island-wide generalisation and state-sponsored ideology.

Diversity, of course, is a double-edged quality, being grounds for both inclusion and exclusion. In his very influential, if deeply pessimistic, analysis of the late nineteenth and early twentieth centuries, the historian F. S. L. Lyons (1979: 177) argues that beside 'the essential unity' of Ireland there is a no less 'essential diversity', 'unbridgeable fissures' deeply embedded in the past and perpetuated by contemporary politics. He sees 'a collision within a small and intimate island of seemingly irreconcilable cultures, unable to live together or to live apart'. The phrase, 'essential unity', was coined by Estyn Evans, who argued from a very different perspective. He depicted the conflict between native and newcomer as being the true dynamic in Irish society, 'the clash that struck the sparks in Irish culture' (1984: 13), Ireland's very insularity attracting invaders and creating the reality of its diversity. His account of the 'personality' of Ireland (1981) argues that the island is no different from the majority of European nations and states which have evolved through a fusion of regional loyalties. The (nine-county) province of Ulster is one strong regional variant within Ireland, if morphologically distinct behind the barrier of difficult drumlin country that stretches from the County Down coast to Donegal Bay (Figure 10. 1). Nevertheless, for Evans, it remains within the essential unity of Ireland, one distinctive element in the island's diversity of habitat, heritage and history (Graham 1994b).

A pluralistic emblematic landscape – with its renegotiation of what is acceptably Irish – includes many of the same places but with different meanings and memories. It also embodies an apparent willingness to accept variant strains of Irish nationality and thus admit a more inclusive landscape in terms of locality and artefacts. For example, the overtly anti-urban nature of Gaelic nationalist historiography was consequent upon the rendition of towns as a central element of a landscape of oppression. The 'environmental revolution' of the eighteenth and early nineteenth centuries, which transformed urban landscapes throughout Ireland to conform to the tastes and values of the landed élite, can now be depicted as one element in a process through which that essentially *arriviste* class set out to affirm its Irishness and claim Ireland as its own (Foster 1988: 191–4; Graham and Proudfoot 1994). Driven by the continuous renegotiation of hegemonic imagery and the ongoing process of nation-building, but also by the economic commodification of the past as tourism, the current Irish myth of place can incorporate these once-excluded artefacts within the canon of its permissible icons (Graham 1994c).

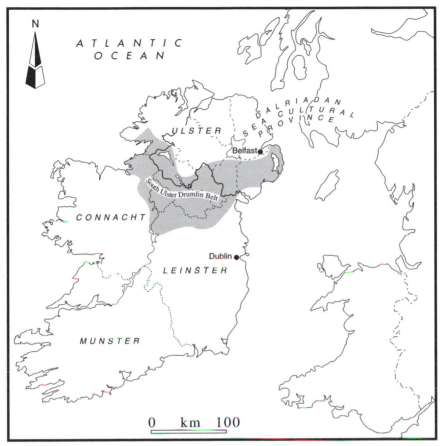

Figure 10. 1 The cultural location of Ulster

THE UNBRIDGEABLE FISSURES?

Identity in Ulster

In many ways, such representations of unity in diversity forged through the renegotiation of landscapes of memory reflect a construction of Irish identity no longer dependent on opposition to the Other for its defining characteristics. To a large extent, this also emphasises the divergent pathways of a Republic of Ireland, redefining and reorienting itself as a modern European state, and Northern Ireland, where unionists and nationalists remain locked into zero-sum thinking on the exclusivity of territoriality, parallel inflexible mind-sets that are apparently oblivious to any conception of the changes repositioning the contemporary Republic. Thus Ulster Protestants remain very sure of what they are not (Catholic Irish), but much less certain about who they are. Superficially, religion may be a far more important element in defining identity for many Protestants than either unionism or Britishness (Pollak 1993), although this may well reflect no more than religion's central function as an ethnic marker (see pp. 130–3).

The increasingly fluid perceptions of Irishness have also had little impact on the representation of Ireland held by many northern nationalists. Like the descendants of the eighteenth- and nineteenth-century diaspora, they may largely continue to subscribe to the traditional discourse in which the six counties of Northern Ireland constitute a temporarily separated part of an inner-directed Irish nation-state, a determinist rendition in which the surrounding sea demarcates the natural national unit. At least superficially a majority of Catholics support the unification of Ireland, the proportion holding this view having increased since the 1994 Downing Street Declaration underlined the lack of British interest in Northern Ireland. The commitment, however, is variable by age and class, younger and middle-class Catholics rejecting both nationalist and unionist labels (Breen 1996). O'Connor's study (1993) also points to major ambiguities in Catholic identity, which can no longer be defined as simply 'Irish'. In part, this reflects processes such as the renegotiation of gender representations and identities (see Chapter 6) and also the strategy of conscious embourgeoisement discussed in Chapter 5, the cleavage of the sectarian axis along gender and class lines having created a dissonance of identity, particularly among middle-class Catholics (Shirlow 1995). According to some commentators, these policies have inverted the rationale of Irish political unification for a socially ascendant Catholic middle class, whose cultural identity may be superseded or diluted by material interests, best served by maintaining strong economic and political links with the UK (for example, Gudgin 1995). Conversely, working-class Catholic attitudes to cultural identity are much more readily – if by no means absolutely – located within the confines of the sectarian discourse.

At an aggregate scale, the contested evidence of the 1991 Census suggests that the ethnic geography of Northern Ireland has become more sharply demarcated, a trend attributable to conflict, more or less voluntary population movement and differential rates of migration (Figure 10. 2). Catholics, who now constitute about 42 per cent of the population, form a substantial majority in all of Counties Fermanagh, Tyrone, parts of Down and Londonderry (including Derry City), together with considerable areas of Belfast. Protestant numerical domination is restricted to the remainder of Belfast, central and north Down, County Antrim (except for its north-east corner) and, finally, the area around Coleraine. Of course, these aggregate patterns conceal both population density and the complex micro-geography of ethnic segregation within the six-county border. It is only at the local scale that Catholics and Protestants occupy spatially discrete and mutually exclusive territories, an intimate geographical proximity which results in

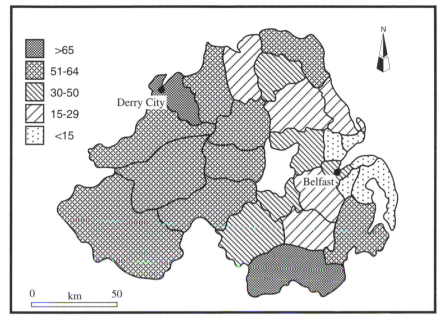

Figure 10. 2 Northern Ireland's sectarian geography
(percentage Catholic in 1991 by local government district)
Source: Northern Ireland Census 1991

relationships between Protestants and Catholics, carefully balanced in peace-time, becoming tense and murderous in crisis – as demonstrated by the sectarian dimension to the Troubles. However, the degree of tension varies with location, class and age.

Integration and devolution – the cultural ambiguity of unionism

In further exploring the issues of identity and place in modern Ulster, the remaining discussion concentrates on unionism, which demonstrates clearly the irreconcilable tensions that ensue from an unagreed representation of place. As Chapter 8 has shown, it is apparent that contemporary unionism is highly fractured. Gallagher (1995), for example, distinguishes between an Ulster Protestant nation and part of the British nation while Porter (1996) – in a brave attempt to construct a theoretical basis for unionism – argues more convincingly that the discourse takes three forms. Cultural unionism (which might be equated to the devolutionist mainstream) relies on its adherence to an exaggerated and heavily qualified perception of the Protestant–British way of life. In contrast, liberal unionism is integrationist, seeking a narrowly political form of freedom and citizenship, which it naively claims to be characteristic of the rest of the UK. Porter rejects both, arguing instead for a civic unionism that recognises both the Britishness and Irishness of Northern Ireland and deals fairly with issues of parity of esteem. Both cultural and liberal forms of unionism are exclusivist because they fail to understand and address nationalist alienation, whereas civic unionism is to some extent congruent with the centrist or accommodationist alignment, already addressed elsewhere (see Chapters 5, 7 and 8), in which cultural and material aspirations are separated. This fracturing of unionism is reflected in a contested set of ill-defined representations of identity which – as Porter argues – interact with political perspectives that encompass the integration of Northern Ireland into the UK and the broad devolutionist perspective that constitutes mainstream unionism but which can merge into advocacy of an independent Northern Ireland (Table 10. 1). Clearly, such political solutions are about judgements related to power as much as identity, but the latter fulfils a key role in legitimating and validating the authority of the former.

Table 10. 1 The fracturing of unionism in Northern Ireland

Political perspective	Alternative labels	Cultural representations
Integration of Northern Ireland into UK state	Liberal unionism Part of the British nation	Denial of Irishness; denial of nationalism; claim to citizenship of multi-national, multi-ethnic UK state.
Devolution	Cultural unionism Ulster Protestant nation Loyalism Ethnic nationalism	Ulsterness defined by location in separate region, culturally and historically distinct from remainder of Ireland but conditionally linked to Britain.
Repartition Independence	As above	As above but with emphasis on internal cohesion rather than dependence on UK allegiance. Ulster is portrayed as a conflict of nationalities.
Accommodationist	Civic unionism Centrist Third force Civic nationalism	Repostioning of Northern Ireland to allow co-existence of both British and Irish aspirations.

The competing versions of unionism have contrasting demands of manipulated geographies. Integrationists – Porter's liberal unionists, Gallagher's 'part of the British nation' – regard Northern Ireland as a physically separate but nevertheless fully integrated region of the UK. To a very considerable extent, the concept of an imagery of identity derived from within the island of Ireland is irrelevant, indeed subversive, to this viewpoint. Northern Ireland, being at one with Britain, does not require any symbolic universe apart from that conferred through its status as a distinct region within a heterogeneous union state. In contrast, devolutionists – adherents of cultural loyalism or the 'Ulster nation' – seek some degree of self-determination, believing Northern Ireland to be more Ulster than it is British or Irish. The union is thus highly conditional, being seen as the most effective means of avoiding incorporation within a Catholic-dominated Ireland and retaining the material benefits accruing from the UK state. The proposal that Northern Ireland be 'cantonised' or repartitioned constitutes a somewhat deviant strand within this broad theme. However, sufficiently extended, and underpinned by a strong sense of grievance stemming from what is now widely perceived as a British betrayal of unionist interests, devolution, in stressing Northern Ireland's cultural and historical separation from the remainder of Ireland, can drift towards the logic of a negotiated independence.

If any form of Irish unity is rejected, and ethnic cleansing of Yugoslavian dimensions abhorred, all strands of unionism – with the exception of integrationism – require some overarching representation that subsumes sectarianism and depicts an integrative, pluralistic myth of Northern Ireland place acceptable to all who live within its disputed boundaries. Only then could unionism as a discourse obtain the legitimacy that would allow it to exercise power over the people who inhabit the territory it seeks to control. Prior to the collapse of the Stormont government in 1972, the unionist leadership displayed a very selective concern with the Province's cultural landscape as an element in the construction of identity. The principal exception was Terence O'Neill (Prime Minister, 1963–9), who – in concert with Estyn Evans – was instrumental in establishing the Ulster Folk Museum at Cultra Manor, County Down, an institution ambitiously designed to demonstrate the cross-sectarian nature of Ulster's rural material culture (Graham 1996). Otherwise, the unionist government adopted a strategy of 'masterly inactivity' toward the past (Clifford 1987), much applauded by Aughey (1995: 15) because it prevented Northern Ireland from becoming a state with illusions of self-determination. However, the unionist leadership, albeit culturally indolent, took great care to ensure that Northern Ireland shared in the material and welfare benefits of the UK state.

Given this studied neglect of the integrative continuities of cultural metaphors, the unionist discourse was reduced to little more than a handful of events (primarily sectarian). Official unionist identity lacks any resonances of a hegemonic, legitimising representation of place but has depended instead on an exaggerated sense of Protestantism, with its history of martyrdom, treachery and Catholic duplicity. Territory can still be claimed through marching and mural but, at the official level, single events set outside place – the 1641 Rebellion (when settlers were massacred by Catholic rebels), the Siege of Derry in 1689, the Battle of the Boyne in 1690, the first day of the Battle of the Somme, the signing of the Anglo-Irish Agreement at Hillsborough in 1985 – have substituted for the communality and continuity embodied in representative landscapes. The iconography of Protestantism, displayed on Orange banner and mural alike, depicts such events in terms of blood sacrifice and/or Catholic deception and, increasingly, as British betrayal. But, crucially, such readings offer neither narrative of continuity nor text of place. Mainstream 'cultural' unionist identity remains largely dependent on depictions of the Other to legitimate its discourse of exclusion.

To integrationists or liberal unionists this does not constitute a dilemma. On the contrary, the lack of an emblematic unionist landscape is a positive virtue, a Britain embodying progressive, liberal and democratic values being the imagined community. Aughey (1989) disputes the widely held perception of unionists as a people in limbo, who have not yet come to terms with Ireland because of narrowly Catholic and aggressively Gaelic versions of Irish identity. He denies that the lack of an unequivocal sense of Protestant national and political identity is a problem. Rather, citizenship of the UK state provides the principle of unity, transcending any need to formulate a distinctive Northern Irish identity which would necessitate accepting the postulates of the nationalist argument and thus equate to a form of embryonic separatism. Consequently, the very idea of a representative landscape as a signification of place is anathema to the integrationist or liberal discourse.

However, this argument, which clearly clashes with Anglo-Irish policy, is also undermined by Ulster's sectarianism (Coulter 1994), no other region of the UK containing a substantial minority population, defined by a strong cultural nationalism and actively seeking unity with another state. Again, Aughey neglects the ubiquitous relationship between political structures and territoriality, the actual legitimacy of any state being defined by the acceptability to its population of the representations of the territory it occupies. Linda Colley (1992: 5–6) argues that Great Britain can plausibly be regarded as an invented nation – forged above all by war – and 'superimposed, if only for a while, onto much older alignments and loyalties'. It is defined not by any domestic political or cultural consensus but in reaction 'to the Other beyond [its] shores' (formerly the French, now the European Commission) (see p. 218). Despite the dismantling of the apparatus of local government by a New Right administration that has transformed the UK into politically the most centralised state in the EU (Hutton 1995), the cultural tensions of this invented nation are readily apparent in the demands for devolution, regional assemblies and cultural recognition for ethnic minorities. The schism within Scotland between those who seek to express their cultural nationalism within a union state, and advocates of political independence, is but one manifestation of the illusory unity of the UK.

This suggests that although unionism could be incorporated within this diverse state in narrowly civic terms, it cannot define itself through assumptions of enduring Britishness when that identity itself is being subjected to radical transformation under the impact of internal and external forces, and is anyway unacceptable to many nationalists in Northern Ireland. Yet there appears to be little or no consciousness within the broad unionist discourse – Porter excepted – of this continual renegotiation of Britishness. In one respect, paralleling their attitudes to territorial sovereignty, British identity is a zero-sum. But in other regards, unionists do profess a very conditional conception of Britishness (Bruce 1994). Mainstream unionism is irredeemably devolutionist, loyalists – the cultural unionists – paradoxically finding their communality within Ulster (McAuley 1994). They may express loyalty or allegiance to the British Crown but, as Miller (1978) persuasively argues, this is a contractual relationship rather than a condition of identity. Logically, if such a covenant is broken, one is released from it. But if that is the case, it underlines the confusions of this axis of unionist identity, which cannot be defined by political criteria if these are negotiable and even ultimately expendable, or by a reactionary rejection of an Other that arguably no longer exists. Furthermore, this insistence upon the conditional nature of the relationship with Britain forces unionists to acknowledge that their identity must somehow embrace their domicile in the island of Ireland and the legitimation conferred to power relationships by a social solidarity fixed in that place. The failure to come to terms with this dilemma lies at the root of the downfall of the unionist polity. The selective and spasmodic history, centred on largely sectarian events, which the unionist state chose for itself, was not only irrelevant to the nationalist population, but also more generally inadequate as a means of legitimisation, principally because of its lack of congruence with place. In the absence of an overarching narrative embracing the mutual reinforcement of power and space, unionists are left lacking legitimacy and authority throughout the territory which they seek to control.

A Common Ground?

All strands of unionism share the belief that the island of Ireland is not a natural socio-geographic entity. In the words of one submission to the Opsahl Commission, 'there is no historical imperative that Ireland should ever be united', Northern Ireland being 'an "unagreed" entity, rather than a non-legitimate one' (Pollak 1993: 17). Unity with what is still widely perceived to be a sectarian Irish state is neither a desirable political nor cultural goal. But, conversely, Northern Ireland cannot survive by looking to a Britain that is at best ambivalent, at worst overtly hostile, and which has declared its lack of political strategic interests in Ireland. Faced with this dilemma, the nationalist rejection of Britishness, and the desire to define some project of legitimation that might transcend sectarianism, the Protestant middle class, acknowledging the notion that Northern Ireland's British allegiance and Protestant ethos are inadequate to define its identity, has often turned culturally to the landscape that, in its indigenous historical and cultural heritage, 'seems on occasion to join the sects' (Foster 1991: 158).

The unifying potential of emblematic landscape as an overt signifier of a common ground between unionist and nationalist has long given rise to discussion. For example, such notions were central to the philosophy of John Hewitt (1907–87) and, indeed, it is in literature, characterised by a sense of place, that the issue of the Province's representative landscape continues to be most comprehensively addressed (see pp. 77–9). Hewitt, a Protestant who could not fall back on religion to define the Irishness in his identity, depicted Ulster as a landscape of singular geographical and economic coherence, one that conveyed a traditional and historical oneness to all its people. In so doing, he directly challenged the unionist assumption that an exploration of Ulster's cultural environment would admit the political mystique of Irishness. Hewitt wanted to invent something quite distinct from the exclusive representative landscapes of Irish-Ireland. But it was to be different, too, from the militaristic Ulster of the new unionism, 'the land of the heroes of the Somme and the generals of England's war' (Vance 1990: 228).

There are, however, alternative and more exclusivist representations of the common ground. While the various devolutionist interpretations of Northern Ireland do not normally question the integrity of the political unit – indeed, they generally seek a cultural underpinning that might strengthen it – the idea of repartition can be regarded as one deviant form of this general perspective. Kennedy (1986) – in an argument that, more recently, has received significant support from elements among the loyalist paramilitaries – contends that the contact between the two political traditions in the 'narrow ground' of Northern Ireland has frequently been mutually deforming. Thus he advocates repartition along ethnic grounds to produce a sustainable unionist Northern Ireland. However, an almost entirely Protestant polity could only be attained through ethnic cleansing, given the intensely localised nature of residential segregation. Nor would it attain any legitimacy in the wider world although, ironically, it would at last fulfil the cultural unionists' assumption that nationalists are not there.

Again, if unionist allegiance is indeed contractual, it follows that another radical common ground might be sought through the abandonment of the union in favour of an independent Northern Ireland state. Recent attempts at delineating an indigenous cultural representation for Northern Ireland emphasise this tension. These aim at creating a northern origin myth of place that would legitimate the claim of Ulster's Protestants to their territory within the island of Ireland (Graham 1994a). Central to such projects, whatever the attitude to the union, is the development of an iconography that emphasises Northern Ireland's cultural separation from the remainder of Ireland. Ostensibly an expression of ethnic nationalism, such propositions are not in themselves sectarian if Catholics are prepared to share in the representations so defined.

The basic premises behind these ideas derive from the assumption of a common past, separated from the remainder of Ireland by the drumlin belt of south Ulster, the most enduring frontier region in the island's history. This cultural distinctiveness is not a product merely of the seventeenth-century plantations, but is rooted in the long-term communality of the Dalriadan Sea cultural province (Adamson 1991) (Figure 10. 1). Ironically, the proponents of this perspective exploit precisely the same sort of heritage sources and artefacts in creating a sense of place as did traditional Irish nationalists in their creation of the Gaelic mythology. Adamson (1978, 1982), for example, seeks to erect a narrative of continuity that links contemporary Ulster to the tales and sagas of the Iron Age. The earliest inhabitants of what is now Ulster controlled an Ulster-Scots cultural province prior to the arrival of the Celts – or Gaels – in Ireland. Gradually, their ancestors were driven back to the area that now constitutes Counties Down and Antrim by Gaelic tribes from the south and west. The hill-fort of Emain Macha (Navan Fort near Armagh), the capital of Ulster during the first centuries BC, becomes the ceremonial centre of this mythology. Cú Chulainn, the hero of the Ulster Cycle, the pseudo-histories of the Province during the Iron Ages, is reincarnated as the leader of Ulster resistance to the invading Gaels.

Consequently, the Scottish migrants who crossed to Ulster in the seventeenth century were the inheritors of a culture, essentially framed in earlier millennia, which linked north-east Ulster, Argyll and Galloway. The Scottish Plantation of Ulster was therefore not a confrontation of alien cultures, nor the oppressive colonialism of the Gaelic myth, but a reunification and reconquest by a Scots-Irish people once expelled from their rightful territory by the invading Gaels. Adamson argues that even the latter, ultimate victors over the Ulaid, had more in common with their ethnic kindred in Scotland than those in the remainder of Ireland. Thus the Scottish – but not the English – Planter and the Gael in north-east Ireland shared a common cultural ancestry. This mythology – blatant even by Irish standards – is being used to construct a representative Ulster discourse of place, a signifying system that communicates and reproduces a separate but ethnically integrated representation of the North, its contemporary stature legitimated by the longevity of its independent and once glorious past. In seeking an authentic expression of Ulster culture that is no longer a pale reflection of those 'psychological colonialisms of Irish nationality and British nationalism' (Foster 1991: 294), this is a populist rather than élitist narrative, much sneered at by professional historians and archaeologists although, as Roy Foster (1989: 4) observes, it does rearrange the pieces in more surprising patterns. It may also be questioned if such ideas have much relevance to many unionists who may be content to see themselves as a settler society, akin to those of North America. Certainly, the whole notion of a Scots-Ulster migratory epic, transcending both the Dalriadan Sea and the Atlantic Ocean, is one deeply entrenched in unionist consciousness.

An indigenous Ulster representation of place – framed in Adamson's terms – is open to exploitation by advocates of both devolution and independence. The tortuous case that the Province has always been distinct and distant from the remainder of Ireland in turn confines the scope of Ulster's British cultural connection to its hypothetical linkages with Scotland. Recent attempts to strengthen these by seeking European Commission recognition of an Ulster-Scots language failed when the EC's Bureau for Lesser Used Languages was unable to find any evidence of a communal language other than English in the Protestant heartland of east Ulster. Indeed, the Ulster-Scots narrative depends on many of the same postulates of separateness from England which have produced a strong Scottish cultural nationalism, albeit contested between those who wish to see that separatism accommodated within the UK state and others who aspire to independence for Scotland within the EU. The logic of an indigenous representation of Ulster points in this latter direction as well, particularly given the lack of reciprocal political support within Scotland for any Ulster-Scots identity. Thus the development of a legitimating and empowering metaphor of place for contemporary Northern Ireland places unionist devolutionists in a double-bind. On one hand, it points to the Irishness within their identity, on the other to the failure of the union to provide anything more substantial than an increasingly compromised political allegiance which is irrelevant to nationalists. Ulster becomes neither Irish nor British, the Ulster-Scots connection demonstrating only that devolution and independence are exclusivist strategies separated by degree rather than kind.

CONCLUSION

The absence of a political consensus within Ireland reflects the contested nature of identity and place discussed here. If any unity of purpose concerning future political structures for the island is to be constructed from the several strands of unionism and nationalism, a cultural environment must first be provided in which ideology creates an integrative place consciousness which, in turn, can signify the holistic and inclusive nature of that philosophy. No matter how impaired the rhetoric might be, the challenge in Ireland is to create cultural landscapes in which inclusive pluralist myths can be embedded. Without the cultural cement of an ordered simplified version of the world, no political framework can achieve legitimacy and Northern Ireland will follow Algeria and South Africa into the history of failed settler societies. There is no far-flung western frontier here, only the 'narrow ground' of six small counties.

It has been argued above that the Irish state is in the process of discarding time-worn representations of place that served to help unify the twenty-six county state (not least by its exploitation of partition) in favour of a sense of place that, while encapsulating the unique qualities of Irishness, is also heterogeneous, outward-oriented, markedly less Catholic and intensely localised. In contrast, Northern Ireland is contested not merely between unionism and nationalism but actually within both camps themselves, which are further riven by class, gender, locality and age divisions. It might be argued that for the middle classes, material prosperity (ironically, much of it directly created by the Troubles) transcends questions of identity but that for the working classes – unionist and nationalist – sectarian consciousness has subsumed class, gender, rural and urban divisions (McAuley 1994: 174–81).

To a very significant extent, this dissonance of identity – ultimately the principal impediment to political negotiations on the future of Ireland – reflects the plethora of places and utter lack of consensus that Northern Ireland has become. Together, many unionists and nationalists espouse a shared insistence on equating cultural identity with territorial sovereignty. Although Ulster Catholics may be British citizens, a majority identify culturally with Ireland. As Scotland shows, cultural nationalism could be incorporated within the UK, albeit with some tensions. However, many Northern nationalists remain locked into the discourse of exclusion that is Sinn Féin's Irish-Ireland, even though this is increasingly divergent from the ongoing renegotiation of place that defines the Republic. In their own ideology of exclusion, cultural unionists (to adopt Porter's term) ignore both revisionism and the simultaneous renegotiation of Britishness, yet possess no agreed alternative hegemonic representation of place to legitimate and validate their cause with all the people of Northern Ireland. Liberal unionism fails to acknowledge the complexity of Britishness, regarding it instead very much as the 'collective social fact' once characteristic of traditional Irish nationalism. It defines the union largely through political criteria while admitting no place for Ireland in Ulster. Faced with the dilemma that any internal cultural synthesis would have to embrace – or at least acknowledge – elements of Irishness, liberal unionism opts to be no more 'than a distant echo of another land' (Foster

1991: 294), one, moreover, that shows little long-term interest in returning its loyalty.

The development of a culturally separate Ulster narrative of place memory, integrated into the Ulster-Scots epic, has been one response to this impasse. However, not only does it embrace connotations of alienation from the union by its imperceptible shading into an argument for independence, but it further depends on the assumption that Ulster is the single distinctly different region in an otherwise geographically homogeneous island. As observed here, contemporary historiography argues against this, depicting a heterogeneous Ireland of many local places in which Ulster, one particular region among a number, is itself extensively differentiated by more parochial loyalties. Thus the unionist predicament remains that, while the development of an indigenous, synthetic, cross-sectarian cultural representation of Ulster is necessary to legitimate power in the terms defined here, inevitably that construct will also support the efficacy of a pluralist depiction of aspirations to Irishness. However, as overlapping and intersecting socio-political networks of power fundamentally redefine Europe, the cultural problem for Ulster unionists remains a vexed one. They cannot continue to say no but must instead formulate the positive cultural iconography necessary to imagine and thereby legitimate their place for all its people. Inevitably, such a construct must take them closer to Ireland but only to a revised and pluralist representation of that society, defined by regional and cultural heterogeneity, notions of hybridity and the equality of rights of citizenship embodied in civic nationalism.

ACKNOWLEDGEMENT

My thanks to Mike Poole for his detailed and perceptive criticisms of an earlier draft of this chapter.

REFERENCES

Adamson, I. (1978) *The Cruthin*, Belfast: Donard.
—— (1982) *The Identity of Ulster*, Bangor: Pretani Press.
—— (1991) *The Ulster People*, Bangor: Pretani Press.
Agnew, J. (1996) 'Time into space: the myth of "backward Italy" in modern Europe', *Time and Society* 5, 1: 27–45.
Ashworth, G. J. (1993) *On Tragedy and Renaissance*, Groningen: Geo Pers.
Aughey, A. (1989) *Under Siege: Ulster Unionism and the Anglo-Irish Agreement*, Belfast: Blackstaff.
—— (1995) 'The idea of the union', in J. W. Foster (ed.) *The Idea of the Union*, Vancouver: Belcouver Press.
Barnes, T. J. and Duncan, J. S. (eds) (1992) *Writing Worlds: Discourse, Text and Metaphor in the Representation of Landscape*, London: Routledge.

Breen, R. (1996) 'Who wants a United Ireland? Constitutional preferences among Catholics and Protestants', in R. Breen, P. Devine and L. Dowds (eds) *Social Attitudes in Northern Ireland*, VI, Belfast: Appletree Press.

Brown, T. (1991) 'British Ireland', in E. Longley (ed.) *Culture in Ireland: Division or Diversity*, Belfast: Institute of Irish Studies.

Bruce, S. (1994) *The Edge of the Union: The Ulster Loyalist Political Vision*, Oxford: Oxford University Press.

Bryson, L. and McCartney, C. (1994) *Clashing Symbols*, Belfast: Institute of Irish Studies.

Clifford, B. (1987) *The Road to Nowhere*, Belfast: Athol Books.

Colley, L. (1992) *Britons: Forging the Nation*, London: Yale University Press.

Coulter, C. (1994) 'Class, ethnicity and political identity in Northern Ireland', *Irish Journal of Sociology* 4, 1–26.

Cullen, L. M. (1988) *The Hidden Ireland: Reassessment of a Concept*, Mullingar: Lilliput Press.

Duffy, P. J. (1994) 'The changing rural landscape, 1750–1850: pictorial evidence', in R. Gillespie and B. P. Kennedy (eds) *Ireland: Art into History*, Dublin: Town House.

Evans, E. E. (1981) *The Personality of Ireland: Habitat, Heritage and History*, Belfast: Blackstaff.

—— (1984) *Ulster: The Common Ground*, Mullingar: Lilliput Press.

Foster, J. W. (1991) *Colonial Consequences: Essays in Irish Literature and Culture*, Dublin: Lilliput Press.

Foster, R. F. (1988) *Modern Ireland, 1600–1972*, Harmondsworth: Allen Lane.

—— (1989) 'Varieties of Irishness', in M. Crozier (ed.) *Cultural Traditions in Northern Ireland*, Belfast: Institute of Irish Studies.

Gallagher, M. (1995) 'How many nations are there in Ireland?', *Ethnic and Racial Studies* 18, 4: 715–39.

Graham, B. J. (1994a) 'No place of the mind: contested Protestant representations of Ulster', *Ecumene* 1, 3: 257–81.

—— (1994b) 'The search for the common ground: Estyn Evans's Ireland', *Transactions Institute of British Geographers* NS 19, 183–201.

—— (1994c) 'Heritage conservation and revisionist nationalism in Ireland', in G. J. Ashworth and P. Larkham (eds) *Building a New Heritage: Tourism, Culture and Identity in the New Europe*, London: Routledge.

—— (1996) 'The contested interpretation of heritage landscapes in Northern Ireland', *International Journal of Heritage Studies* 2, 1 and 2: 10–22.

Graham, B. J. and Proudfoot, L. J. (eds) (1993) *An Historical Geography of Ireland*, London: Academic Press.

—— (1994) *Urban Improvement in Provincial Ireland, 1700–1840*, Athlone: Group for the Study of Irish Historic Settlement.

Gruffudd, P. (1995) 'Remaking Wales: nation-building and the geographical imagination, 1925–50', *Political Geography* 14, 3: 219–39.

Gudgin, G. (1995) 'The economics of the union', in J. W. Foster (ed.) *The Idea of the Union*, Vancouver: Belcouver Press.

Heffernan, M. (1995) 'For ever England: the Western front and the politics of remembrance in Britain', *Ecumene* 2, 3: 293–324.

Hutton, W. (1995) *The State We're In*, London: Jonathan Cape.

Jarman, N. (1992) 'Troubled images', *Critique of Anthropology* 12: 179–91.

Jarman, N. and Bryan, D. (1996) *Parade and Protest: A Discussion of Parading Disputes in Northern Ireland*, University of Ulster, Coleraine: Centre for the Study of Conflict.

Johnson, N. (1994) 'Sculpting heroic histories: celebrating the centenary of the 1798 rebellion in Ireland', *Transactions Institute of British Geographers* NS 19: 78–93.

—— (1995) 'Cast in stone: monuments, geography and nationalism', *Environment and Planning D: Society and Space* 13: 51–65.

Kennedy, L. (1986) *Two Ulsters: A Case for Repartition*, Belfast: the author.

Lee, J. J. (1989) *Ireland, 1912–1985: Politics and Society*, Cambridge: Cambridge University Press.

Leonard, J. (1996) 'The twinge of memory: Armistice Day and Remembrance Sunday in Dublin since 1919', in R. English and G. Walker (eds) *Unionism in Modern Ireland*, Dublin: Gill and Macmillan.

Longley, E. (1991) 'The rising, the Somme and Irish memory', in M. Ni Dhonnchadha and T. Dorgan (eds) *Revising the Rising*, Derry: Field Day.

Lyons, F. S. L. (1979) *Culture and Anarchy in Ireland, 1890–1939*, Oxford: Oxford University Press.

McAuley, J. W. (1994) *The Politics of Identity: A Loyalist Community in Belfast*, Aldershot: Avebury.

MacLaughlin, J. (1993) 'Place, politics and culture in nation-building Ulster: constructing nationalist hegemony in post-Famine Donegal', *Canadian Review of Studies in Nationalism* XX: 97–111.

Miller, D. (1978) *Queen's Rebels: Ulster Loyalism in Historical Perspective*, Dublin: Gill and Macmillan.

O'Connor, F. (1993) *In Search of a State: Catholics in Northern Ireland*, Belfast: Blackstaff.

Officer, D. (1995) 'Representing war: the Somme Heritage Centre', *History Ireland* 3, 1: 38–42.

Ó Tuathaigh, G. (1991) 'The Irish-Ireland idea: rationale and relevance', in E. Longley (ed.) *Culture in Ireland: Division or Diversity*, Belfast: Institute of Irish Studies.

Pollak, A. (ed.) (1993) *A Citizens' Enquiry: The Opsahl Report on Northern Ireland*, Dublin: Lilliput Press.

Porter, N. (1996) *Rethinking Unionism: An Alternative Vision for Northern Ireland*, Belfast: Blackstaff.

Rolston, B. (1991) *Politics and Painting: Murals and Conflict in Northern Ireland*, Toronto: Fairleigh Dickenson University Press.

Samuel, R. (1995) *Theatres of Memory*, 1, *Past and Present in Contemporary Culture*, London: Verso.

Shirlow, P. (ed.) (1995) *Development Ireland: Contemporary Issues*, London: Pluto Press.

Stewart, A. T. Q. (1977) *The Narrow Ground: Aspects of Ulster, 1609–1969*, London: Faber.

Tunbridge, J. E. and Ashworth, G. J. (1996) *Dissonant Heritage: The Management of the Past as a Resource in Conflict*, Chichester: John Wiley.

Vance, N. (1990) *Irish Literature: A Social History*, Oxford: Basil Blackwell.

Whelan, K. (1992) 'The power of place', *Irish Review* 12: 13–20.

—— (1993) 'The bases of regionalism', in P. Ó Drisceoil (ed.) *Culture in Ireland: Regions, Identity and Power*, Belfast: Institute of Irish Studies.

Part IV

PLACE, IDENTITY AND POLITICS

INTRODUCTION

This book has argued that societies are constituted of multiple, overlapping and intersecting socio-spatial networks of places and axes of identity. The boundaries of these axes rarely overlap and people may occupy conflicting locations within them. If intransigent unionism continues to depend on a nineteenth-century sectarian discourse, largely defined by an equally partisan republican ideology which is now rapidly being deconstructed along the axes of identity considered in this book, it is left in an extremely vulnerable position in responding to the continuing transformation of Irish, British and European society. Nor do these inevitably point to a united Ireland. Although political solutions seek structures of convergence, contemporary Ireland is in many respects characterised by the divergence of its constituent parts, largely because cultural transformation in the North has been restricted to renegotiations of gender and the even more ambiguous creation of an enlarged and partly desectarianised middle class which is less antipathetic to all-Ireland institutions.

Clearly a hybrid, diverse, pluralistic and open-ended conception of Irishness contains the potential to transform Ireland into many Irelands. The implications of these changes have yet to be addressed either by unionists or northern nationalists, but they are fundamental to any future political solution. In the single chapter in Part IV, James Anderson offers a conclusion to the recastings of Irish identity which have been discussed in the previous chapters. He argues that in this 'small and diverse island' the overlapping and multiple dimensions to identity – transnational, national and subnational – analysed here demand a relaxation of the archaic British conception of territorial sovereignty as an exclusive and indivisible absolute. Both an all-Ireland Republic and a British Northern Ireland are 'mutually unattainable bargaining positions'. This book suggests that Ireland is moving away from ethnic conceptualisations of identity towards a civic, participatory democracy within the European Union in which identity is defined by a complex of non-national interests and practices. It is in that direction that flexible political institutions – which do not equate sovereignty with exclusive territoriality – must follow.

213

11

TERRITORIAL SOVEREIGNTY AND POLITICAL IDENTITY

National problems, transnational solutions?

James Anderson

INTRODUCTION

Territoriality and sovereignty are central to the disputed question of political identity in Ireland. They underlie a conflict which is variously seen as 'the Irish national problem', 'the Northern Ireland problem', or 'Ireland's British problem', depending on one's perspective. And whatever the viewpoint, most observers would probably agree that some recastings of territorial sovereignty and identity are essential if the problem is to be solved. They disagree, however, over how and by whom such restructuring might be achieved. There is also disagreement as to the meaning of sovereignty and whether it is still relevant at the end of the twentieth century. Globalisation and the European Union are in some respects transcending exclusive forms of state authority and politics based on national territory. The prospects for a significant recasting of exclusive territorialities in Ireland through North–South institutions seem better than in the 1920s, or even the 1970s, when previous attempts to bridge the border with a Council of Ireland failed.

This three-part chapter – which draws extensively on earlier work (Anderson 1994, 1996; Anderson and Goodman 1994, 1995) – assesses the implications of intensified globalisation and European integration for Ireland's conflicting national identities and North–South relations. First, it examines some of the problems of nationalism, exclusive territoriality and the nation-state ideal, and their particular legacy in Ireland. Second, I discuss the limited impact on these problems of the transnationalising trends and sovereignty changes of recent decades. Despite hopes dating from the 1950s, Europeanisation has so far failed to deliver a solution; and there is little evidence that national identities are being superseded by a European one – in fact the very processes of transnational integration can reinvigorate traditional nationalisms. Finally, while notions of post-nationalism are wishful thinking, it is perhaps even more misleading to insist that nothing has changed. New possibilities are indeed being opened up for a settlement based on North–South institutions. Could Ireland, a byword for supposedly atavistic national problems, be the harbinger of new transnational solutions? As argued here, actually establishing these institutions and settling the national conflict would require new policies and also mobilisation around *non*-national identities and issues which have generally been crowded out by the all-consuming nationalisms.

NATIONAL PROBLEMS

Nationalism's tragic ideal

Northern Ireland is the residue of failures in nation- and state-building, whether viewed from either a British or an Irish perspective. The disputed labels for territory – repeated at a local level in, for example, Derry/Londonderry (or, ironically, 'Stroke City') – reflect the rival national identities and suggest that the ideal of the nation-state is unachievable. Nationalism developed historically in close association with the rise of the modern state and territorial sovereignty. The long medieval-to-modern transition involved a territorialisation of politics, with a sharpening of differences at the borders of states and of nations, between internal and external, belonging and not belonging, us and them. The nested hierarchies and multiple levels of authority in medieval Europe, with overlapping sovereignties defined in terms of functional obligations as well as in loosely territorial terms, gave way to sovereignty delimited only and much more precisely by territory. In effect the multilevel medieval authorities were collapsed to one all-important level, that of the sovereign territorial state, as authority within the territory was centralised and outside powers were excluded. Formal sovereignty over everything – secular and spiritual – was bundled together into territorial parcels called states and, later, nation-states.

This state territorialisation was a precondition for the doctrine of nationalism which links historically and culturally defined territorial communities – nations – to political statehood, either as a reality or as an aspiration. As explained in Chapter 1, nations and states are entities that explicitly claim, and are based on, particular geographical territories, as distinct from merely occupying geographical space which is true of all social activity (Anderson 1986: 117). The nationalist ideal is that they should coincide geographically in *nation-states*: the nation's territory and the state's territory should be as one, each nation having its own state, and each state expressing the general will of a single, culturally unified nation.

The ideal of each nation freely exercising its right to self-determination in its own sovereign, independent nation-state has a strong democratic, popular appeal. But nationalism promises much more than it can deliver. In practice, nations and states rarely coincide, and in many cases they leave sizeable

216

national minorities on the wrong side of state borders. The happy spatial coincidence of cultural community and political sovereignty is rarely achieved in reality, and attempts to make reality fit the ideal have often had unhappy, indeed tragic, consequences. Nationalism has been implicated in some of the twentieth century's worst atrocities, the ethnic cleansing in former Yugoslavia being just one recent example. Even where the ideal of geographical coincidence is approximated, democracy is often sadly lacking. Nationalism is two-faced in several respects: forward-looking but also back-ward-looking to an often mythical or invented past; and divisive at the same time as it is unifying. It brings together different groups and classes in a political-cultural community defined as the people or nation, but it simultaneously separates out different peoples, thus fuelling conflicts between nations and between states or, at the very least, impeding transnational interdependencies and co-operation. And, as Chapter 9 has suggested, the limited unity it offers around the national interest often serves the interests of dominant social groups and classes, rather than the whole nation. Substate national conflicts, such as those in Yugoslavia and Ireland, are essentially about national sovereignty defined in traditional territorial terms. But in situations where people with conflicting national allegiances are intermingled in the same territory, their conflicts are likely to lead to problems of political deadlock, or violence, or both. Hence the attractions of redefining sovereignty and territoriality – but also the dangers of wishful thinking.

State- and nation-building failures

State-building and nation-building developed very unevenly over time and space, and often in direct conflict with each other. Thus in some states – and the UK is a spectacularly good example – nation-building by the state to create a single nation has been confronted by non-state and opposing nationalisms with their own aspirations to separate statehood – in Scotland, Wales and especially Ireland. Irish and British nationalisms and associated state forms developed historically in close mutual opposition, though from very different origins and in very unequal, contrasting ways.

The state-building of England's monarchs met its most serious obstacles in an Ireland which, unlike Britain, remained largely Roman Catholic – a potential ally for England's main Catholic rivals, Spain and France, to whom Irish opponents of British rule periodically looked for help. By the eighteenth century, landownership in Ireland was monopolised by an episcopalian landed élite, which instituted the Penal Laws discriminating against Catholics, and also against non-episcopalian Protestant Dissenters, mainly Presbyterians. But it was only when influenced by the French Revolution that the resulting Irish discontents came to be expressed in nationalism and republicanism. Ireland's first popular nationalist movement, the 'Society of United Irishmen' – committed to republicanism, to 'breaking the connection with England, the never-failing source of all our troubles', and to the explicitly anti-sectarian objective of replacing 'Catholic, Protestant and Dissenter with the common name of Irishman' – was mainly initiated by Belfast Presbyterians and established in 1791. Four years later the landed class responded by sponsoring the Protestant Orange Order. The United Irishmen were militarily defeated in 1798 after the expected French help failed to arrive, while Ireland was fully incorporated into the United Kingdom of Great Britain and Ireland in 1800, largely because of British fears that another French invasion might prove successful.

State-sponsored British nationalism was, according to its leading historian Linda Colley, 'heavily dependent . . . on a broadly Protestant culture, on the threat and tonic of recurrent war, particularly war with France, and on the triumphs, profits and Otherness represented by a massive overseas empire' (Colley 1992: 6). A series of wars with Catholic France over the one hundred and thirty years up to 1815, and associated popular anti-Catholicism, enabled the superimposition of an imperialistic British identity on separate national and local identities. The multi-nation-state, including Ireland, was held together by an archaic conception of sovereignty as the absolute and indivisible preserve of the London-based Parliament and Crown.

But Ireland, more Catholic than Protestant, 'was never able or willing to play a satisfactory part' in this Britishness:

> Cut off from Great Britain by the sea . . . it was cut off still more effectively by the prejudices of the English, Welsh and Scots, and by the self-image of the bulk of the Irish themselves, both Protestants and Catholics.
>
> (Colley 1992: 8, 322–3)

By contrast, Irish nationalism, although in practice often imbued with Catholicism, is anti-sectarian in principle as well as origin, and for most of its life it has been an oppositional and substate movement in a British-dominated context. It still is in Northern Ireland, the present apex of state-building and nation-building failures, where British and Irish nationalisms now meet head-on in tragic testimony to the nation-state ideal.

Northern Ireland was the outcome of a failed unionist attempt to prevent Irish nationalists achieving Home Rule, and Ireland's partition represented a retreat for British state-building. However, as the previous chapter has shown, Ireland's separatist nationalism failed in – indeed was incapable of – securing the allegiance of a majority of the predominantly Protestant population of north-east Ireland (ironically the area of its main founders). Partition meant that part of the claimed national territory, together with a disaffected Irish nationalist minority remained inside the British state. Unionists, basing themselves mainly on the nine-county province of Ulster – 'Ulster will fight, and Ulster will be right' – had campaigned under the slogan of 'Home Rule is Rome rule' to block legislation for the whole country. In 1886, the Westminster House of Commons voted in favour of Home Rule, only to be over-ruled by the unelected House of Lords. But when it became clear in the decade before 1920 that this tactic could not succeed, the Ulster unionists opted for a six-county Northern Ireland to give themselves a 'safe', roughly two to one, majority of Protestants, assumed to be unionists. They would have had a much narrower majority (between 45 and 55 per cent) if the three Ulster counties with large Catholic majorities – Donegal, Cavan and Monaghan – had not been excluded.

The subsequent conflict has sometimes been termed a double minority problem. While nationalists now comprise a substantial minority of Northern Ireland's 1.5 million population, unionists are outnumbered by more than 4:1 in Ireland's total population of some five million. The conflict might, however, be better seen as a quadruple minority problem. If we take the UK as the territorial unit, Northern unionists constitute less than 3 per cent of a population in excess of 55 million, potentially vulnerable to a majority of the other 97 per cent which is not committed to Northern Ireland remaining in the UK. So the North's unionists might be seen as vulnerable on two fronts – or as the 'tail' which has so far succeeded in 'wagging' the Irish and the British 'dogs'. Partition also produced a now almost forgotten small unionist minority of about 10 per cent in the South, including some particularly embittered Ulster unionists in the three excluded counties, who felt as much abandoned by their 'own side' as Northern nationalists did by theirs (see pp. 138–9).

Contemporary deadlock

Preserving Northern Ireland's built-in Protestant majority has been a unionist imperative so territoriality, defined in sectarian religious terms, is a key issue, particularly as the present Catholic minority of around 42 per cent represents a significant increase from about 35 per cent in 1971 (Anderson and Shuttleworth 1994). Territoriality, 'a spatial strategy to affect, influence, or control resources and people, by controlling area' (Sack 1986: 21), is linked across different local and national spatial scales in Northern Ireland. The contested state territoriality and sovereignty gives meaning and virulence to local territorial conflicts; local territoriality is used as a metaphor for political dominance or resistance at the level of the state; and local conflicts are seen as contributing to the maintenance (or removal) of the perhaps not-so-safe unionist majority in Northern Ireland as a whole. The long-established tradition of unionist Orange marches through predominantly Irish nationalist localities asserts that these localities are part of the British state territory, rather than belonging to their local nationalist inhabitants. On the other hand, predominantly unionist localities are often seen by their inhabitants and others as unionist territory. It seems that strong local senses of places, in the plural, largely based on sectarian territorialities, help militate against a single myth of Northern Ireland territory as a whole. Such a myth could be internally unifying, whereas myths based on religious difference cannot.

It is because they lack a specific Northern Ireland nationalism and a secular myth of place (as Chapter 10 argued), that unionists mobilise around exclusively Protestant institutions like the Orange Order. The unionism of Northern Ireland can be seen as a very particular – and increasingly distinct and separated – strand of British nationalism. Shaped less by concerns with Empire or France, and more by its own conflict with Irish nationalism, mainstream unionism continues to rely on anti-Catholicism and is irredeemably sectarian. For its part, the traditional overarching British identity is in decline, with the weakening of its formative influences such as the Empire and Protestantism (Colley 1992: 8). Recently, as we have seen (see pp. 207–8), some unionists have attempted to invent a brand new Northern Ireland nationalism – a rare occurrence in late twentieth-century Europe – but with little success because such a specific, separate nationalism would run completely counter to Northern Ireland's *raison d'être* of maintaining the union with Britain. Britishness for a majority of the predominantly Protestant unionists was, and still is, a promise of sectarian advantage over local Catholics and a bulwark against incorporation as a religious minority in an all-Ireland state.

Unionist self-identification as British has in fact increased significantly since the present Troubles started in the late 1960s, a further failure for Irish nation-building and some evidence of the counter-productive aspect of the IRA's military campaign. Trew (1996) shows that whereas Northern Catholics mainly identified themselves as 'Irish' (62 per cent), 'Northern Irish' (28 per cent), or 'British' (10 per cent) in 1994, Protestants mainly saw themselves as 'British' (71 per cent), 'Ulster people' (11 per cent), or 'Northern Irish' (15 per cent). Comparison with a 1968 survey showed that people of Protestant background describing themselves as 'British' grew from 39 to 71 per cent, while those describing themselves as 'Irish' declined from 20 to only 3 per cent, though over twice that number, 7 per cent, favoured Ireland's reunification (Trew 1996: 141–2, 149). The relationship between national identity and political ideology was not totally clear-cut, over a third of respondents considering themselves neither unionist nor nationalist. But in 1994, 27 per cent of all respondents – and 60 per cent of Catholics – favoured reunification. 'Remaining part of the UK' was the preference of 63 per cent overall – and of over 90 per cent of Protestants and 24 per cent of Catholics, though a gradual increase in Catholic support for the union was sharply reversed in 1994 (Breen 1996: 34–6).

The deadlocked nature of the conflict is directly related to the centrality of sectarian territoriality, and the peculiarly archaic British conception of sovereignty as an exclusive and indivisible territorial absolute. Both encourage a zero-sum mentality which precludes any solution short of the all-out but unattainable victories both sides have traditionally sought. The total amount of territory to be divided between them is clearly fixed and more for one side does mean less for the other. But economic and social developments do not have a fixed total, and here the zero-sum approach is very misleading. In fact, the supposed zero-sum game is really a negative-sum game in which the majority on all sides lose. The unionist assertion of exclusively British territoriality may be a pyrrhic victory, the financial costs of which are mainly borne by the increasingly alienated taxpayers of Britain. In theory the unionists have a 'winner takes all' form of sovereignty, but most of the supposed winners are actually losers, with working-class Protestants as well as Catholics bearing the brunt of the conflict.

For unionists to insist on a purely internal settlement is to insist on retaining intact the existing territorial framework of the six counties which, as we have seen, is the macrocosm which mutually exacerbates microcosms of local conflict. Worse still, the continuing failure of government attempts to start meaningful political negotiations about the macrocosm's future means that local conflicts continue to function as a surrogate; if allowed to continue, this could result in attempts at cantonisation or repartition. Given the geographical intermingling of unionists and nationalists this could well lead to so-called ethnic cleansing on a hitherto unprecedented scale.

The conflict is really about which is the appropriate territorial decision-making unit and electorate – the North alone, North and South together, or the whole UK, or even the UK plus the Irish Republic? To decide the electorate is basically to decide its majority decision, and a good case could be made for directly involving all three electorates, North, South and Britain, for all are adversely affected by the conflict. But despite arguments over democracy in Northern Ireland, and whether its delimitation as just six of Ulster's nine counties was an undemocratic gerrymander, the conflict is not susceptible to conventional democratic resolution, precisely because it is fundamentally about who should have a vote, and which state body or bodies should organise the elections or referendums in the first place. Herein lies the intractability of the conflict – and the attractions of redefining territorial sovereignty. If the non-coinciding reality of nations and states is a tragic disappointment, and making reality fit the ideal is either impossible or not worth the cost, perhaps it is the ideal that should be changed?

TRANSNATIONALISING TRENDS

Globalisation and Europeanisation

Intensified globalisation and, more particularly, European integration are in fact redefining sovereignty, though whether this will produce transnational solutions to national problems such as Ireland's is impossible to predict. However, the partial erosion of exclusive territorialities in the European Union is at least increasing the chances of a settlement based on North–South institutions bridging exclusive sovereignties. The nation-state ideal as a guide to political action has become more deficient with globalisation and EU integration over the last two decades. The search for alternatives is at once more pressing and more plausible. There seems even less reason to pursue an unachievable ideal as it now hampers political rearrangements which would be more functionally suited to the realities of our increasingly globalised world. However, functionalist post-national outcomes are unlikely, given the contradictions of globalisation which can actively stimulate nationalism at the same time as calling for transnationalism. John Hume, leader of the Social Democratic and Labour Party, has argued that the nation-state is outdated and that what is important are people rather than territory. With others (Kearney 1988), he has talked of a Europe of the regions replacing the Europe of nation-states. The implication is that national sovereignty and nationalist conflict are being rendered historically redundant and that the North and South of Ireland can come together harmoniously as two regions of a federal Europe. However, unionists, not surprisingly, are suspicious of a post-national regionalism which just happens to deliver the traditional demand of Irish nationalism, while opting for people rather than territory could on the face of it be a spurious choice, for the two are not unconnected. John Hume may believe that 'the day of the nation-state is dead and gone', 'but I haven't had mine yet', retorts Bernadette McAliskey, a former Northern MP and prominent socialist republican (O'Connor 1993: 371).

To indulge in wishful thinking about the wider context – glib notions of a borderless world, an end of territorially based sovereignty, or a European identity replacing national ones – is actively misleading. Rumours about the death of the nation-state are greatly exaggerated (Anderson 1995). New thinking is certainly needed but it will have to be more discriminating, less apocalyptic. Nationalism is very much alive, as indicated by the current rash of national problems across the globe; in some respects the same old territorial states with their sovereignty defined by the same old borders seem as firmly rooted as ever. On the other hand, as argued in more detail elsewhere (Anderson 1996), globalisation is shifting the ground under established political arrangements and concepts, changing the political stage as well as the actors. John Hume's political rhetoric may exaggerate but he is tapping into an emerging European reality. To cling, as unionists do, to the realist view of international relations and the belief that state sovereignty is sacrosanct is even more self-deluding. Clearly, we need to steer a course between the misleading extremes, but how should transnationalisation be characterised?

New territorialities in the European Union

Transnationalisation has developed furthest in the EU, especially since the advent of the Single European Market (SEM) in the late 1980s. It possesses all the transnational, functional and often non-territorial institutional frameworks and regulatory regimes that have recently mushroomed across the world (McGrew 1995). But in addition, the EU is developing supra-state institutions which have been gaining significant transfers of sovereignty from the member states. Some of these institutions, particularly the European Commission, are strengthening substate regionalism, as is the SEM. They are accentuating regional diversity within states, and encouraging a more fine-grained region-to-region integration across state borders, including the Irish border, as distinct from simply state-to-state co-operation (Anderson and Goodman 1995).

The EU has been characterised as perhaps the world's 'first truly postmodern international political form', distinct from the national and federal state forms of the modern era, but in some respects reminiscent of premodern territorialities (Ruggie 1993). This is in line with the hypothesis of a 'new medievalism' – the speculation that the growth of transnational corporations and networks, combined with a regional integration of states as in the EU and a disintegration of states because of substate nationalist and regionalist pressures, might produce overlapping forms of sovereignty analogous to the complex political arrangements of medieval Europe (Bull 1977: 254–5). Sovereignty, rather than being monopolised by states, would again be shared between different institutions at different levels, some based on bounded spaces, others defined more in non-territorial or functional terms (like papal authority in medieval England, for instance, before Henry VIII territorialised spiritual as well as secular sovereignty). A return to overlapping or segmented authority is most likely, not where states die or are replaced by a federal Europe of regions, but where the changes are more partial and ambiguous, undermining but not relocating sovereignty as presently understood.

The basic argument is that the premodern to modern territorialisation of politics, with sovereignty over everything being bundled into territorial state parcels, was associated with what Harvey (1989: 242) has called 'a radical reconstruction of views of space and time'. Conversely, in the contemporary period, global 'time–space compression' is again radically reconstructing our views of space, and leading to an accelerated unbundling of territorial sovereignty, with the growth of common markets and various transnational functional regimes and political communities not delimited primarily in territorial terms (Anderson 1996). De-territorialisation and unbundling may be the key to understanding the contemporary spatial reorganisation of politics.

However, a number of qualifications are necessary before discussing the possible implications for Ireland. First, even in the EU, the unbundling is limited and partial, affecting different state activities very unevenly. The politics of economic development is the sphere where state power has been most affected by globalisation, and it is also generally the main focus for the growth of regional and city politics, as local areas strive to attract external investment capital and avoid peripheralisation. But in some aspects of social and environmental policy, for example, the powers and involvement of the state are as great as ever, and in some cases are still increasing. While territoriality is becoming less important in some fields (for example, financial markets), for many aspects of social, cultural and indeed political life, the state is still the main spatial container. Second, although a new political form, the EU itself is still territorial, and in many respects traditional conceptions of sovereignty remain dominant, whether exercised by the member states or by the EU as a whole. Sometimes there is re-territorialisation rather than de-territorialisation. Third, the EU's democratic deficit is at least partly due to the diffuseness of its shared or overlapping sovereignty and the powerlessness of its central parliament. 'New medieval' analogies have serious limitations, especially with respect to popular democracy and nationalism, both notably absent in medieval Europe.

Thus the foreseeable political reality is likely to be a complex mixture of conventional and new or hybrid forms, with territorial and non-territorial types of community and authority coexisting and interacting. States may well remain the most important political institutions, but they are increasingly having to share the world stage with other new, or newly important, international actors, including a growing number and variety of transnational, functionally defined institutions and movements. Territorially based sovereignty is not ending, but territory is losing some of its importance as the basis of political decision-making. The case against exclusive territorialities has been substantially strengthened.

North and South in the European Union

The main implications of these processes for Ireland could lie in the improved possibilities of escaping zero- and negative-sum games. The potential losses, or gains foregone, from failing to do so, are being increased by integration in the EU, Europe having introduced an important new economic dynamic into North–South relations. This, in combination with socio-cultural parity for Northern nationalists and unionists, could conceivably facilitate a political settlement. But, again, some qualifications are in order. For instance, the UK's and Irish Republic's joint membership of the Common Market and the European Economic Community did not dissolve the national conflict – or the Irish border – as some Irish nationalists had been expecting since as long ago as the 1950s (Anderson 1994). Initially, indeed, it led to greater divergence between North and South as the two states reacted very differently to European integration. Quite conceivably, it could do so again, depending, for instance, on how London and Dublin respond to monetary union. The dashed hopes that Europe would be the catalyst to solve or supersede the national question were in fact grossly inflated, if not an excuse for a lack of proactive policies.

Nor have national identities in the North been noticeably changed, never mind superseded, by EU membership. On the contrary, many EU issues have been largely sectarianised as just another platform for the national conflict. However, the EU's potential for a non-neutral impact on the conflict is reflected (albeit perhaps exaggeratedly) in the sharply differing local attitudes to it. As Trew (1996) reports, whereas in Britain social class, education and age are important predictors of attitudes to Europe, political allegiance is the crucial factor in Northern Ireland. Unionist supporters (like Britain's right-wing Tories) often oppose a European federal superstate as a threat to traditional British sovereignty, whereas John Hume's SDLP supporters tend to see European integration as promising a solution to the sovereignty problem. While 73 per cent of the respondents who self-identified as Irish wanted the UK to be fully integrated into the EU, 57 per cent of those self-identifying as British wanted the UK to maintain its independence from the EU. In terms of religion, 62 per cent of Catholics wanted closer links with the EU, compared to only 32 per cent of Protestants. However, 45 per cent of Protestants were happy with the existing extent of integration, compared to only 34 per cent of the more Europhobic population of Britain (Trew 1996: 145–7) – a difference which probably reflects the fact that Northern Ireland, unlike Britain, is a net beneficiary of EU financial transfers.

Despite the sectarianisation of EU issues and the dashing of past expectations, the present hopes (and fears) vested in Europe are not entirely fanciful. In some important respects it is a new situation. The SEM and associated political developments are much more integrative than previous initiatives. The unionist adherence to the traditional British conception of absolute sovereignty within the state's frontiers does not sit easily with the current reality of shared sovereignty across the EU. Its institutions not only constitute the present context but also provide models for possible North–South institutions in Ireland, although the appropriateness of these exemplars is sharply contested. So too are suggestions that the EU's central institutions should play a more direct role in a settlement, or in pressurising the British government – as has happened in the past. According to former Irish Taoiseach, Dr Garret FitzGerald, joint membership of EC bodies greatly facilitated working relationships between Irish and British personnel, particularly in formulating the 1985 Anglo-Irish Agreement (AIA) (though this fact then insulated the British government from further European pressure).

Nevertheless, if not openly admitted by the British government, constitutional developments in Northern Ireland already impinge on a strict definition of absolute and indivisible Westminster sovereignty. A majority in Northern Ireland could now hypothetically take the region out of the UK if it so wished, a right not granted, for example, to Scotland or Wales. Westminster orthodoxy is also breached when Irish nationalists and unionists lobby external authorities for support, whether it is the Irish Republic, the EU or the USA. The 1985 AIA gave another state, the Irish Republic, a limited consultative role in the affairs of Northern Ireland. Furthermore, full British–Irish sovereignty in the sense of joint London–Dublin authority over Northern Ireland has been considered as one policy option by the British Labour Party. As we shall see, some such extension of the AIA remains a possible option if unionists are not persuaded to agree to direct North–South (i.e., Belfast–Dublin) institutional arrangements.

There have also been novel suggestions that Irish national identity can be separated from the traditional territorial aspiration to a united Ireland, in the sense that the Irish cultural identity of Northern nationalists could be accommodated in Northern Ireland without changing statehood or political sovereignty. But those unionists who would accept that (for example, Cadogan Group 1992) would not agree that a similar separation of cultural identity from political identity and state power could equally apply to unionists. On the other hand, O'Connor (1993: 44–6) found that the political allegiances of many Northern Catholics are 'shifting, complex and ambiguous', and some reportedly did want 'parity of esteem' for an Irish identity within the UK state – within Northern Ireland – albeit backed up by political structures linking North and South.

The Single European Market

More tangible evidence of transnationalisation is to be found in the SEM. The threat of Ireland's further peripheralisation, due to stiffer competition from the stronger continental economies in the SEM, has led to widespread calls for greater political and economic co-operation between North and South. The threat is likely to increase as the EU expands eastwards, and it is especially serious for the weaker Northern economy with its chronic dependency on the massive subvention from Britain's taxpayers. On both sides of industry, in both parts of Ireland, minds have been concentrated wonderfully on the need for economic co-operation, a pooling of resources and policy co-ordination across the island.

Especially noteworthy is the fact that business people of unionist background, who in the past might have shown little interest in the South or looked down on it as backward, now support the call for a single island economy. In 1992, for example, the year the SEM was officially completed, George Quigley, a leading banker and then head of the Northern Ireland branch of the Institute of Directors, proposed that 'Ireland, North and

South, should become one integrated "island economy" in the context of the Single European Market'. The unified economy should be supported by a special EU fund for projects agreed by the British and Irish governments, together with the European Commission, thereby ensuring a direct route to Brussels for a Northern Ireland administration if powers were devolved from London (cited in Anderson 1994: 59).

Because business leaders, North and South, clearly see economic integration as a desirable end in itself, rather than a means to an end, they adopt a resolutely non-political posture in distancing themselves from the nationalist objective of a politically united Ireland. Nevertheless, and not surprisingly, they have been attacked by the unionist leadership, causing a further fracturing of the traditionally close links between the main unionist party and Northern business interests. Clearly, the integration of Northern and Southern economies does have profound political implications, despite its economic motivation and the non-political posturing that business people feel compelled to adopt. Private business interests cannot achieve economic integration on their own, not least because of the importance of the public sector and state involvement in the two economies. Integrating the economies requires North–South institutions to give the process coherence and to provide democratic accountability in North–South policy-making.

The SEM thus constitutes a major new dynamic in relations between the two states, and between North and South as regions of the EU. Although basically an economic dynamic, it is probably the most important new element in the deadlocked conflict. This is especially so because the economic imperatives of the SEM coincide with the quite separate political objective of achieving parity of esteem in the North, in that both require political institutions linking North and South. Together they provide the basis for the settlement envisaged by the two governments in their 1995 Framework Document (HMSO 1995).

NORTH–SOUTH INSTITUTIONS

The joint Framework Document sees parity of esteem for Northern nationalists as requiring a significant Irish dimension in the politics of Northern Ireland to combine with the existing British dimension. It recognises that Northern nationalists do need practical recognition of their identity in the form of institutional linkages with the South. The two governments also proposed an island-wide Parliamentary Forum where unionists and nationalists could 'acknowledge their respective identities', together with an island-wide civil rights 'Charter' to protect the rights of nationalists and unionists, Protestants and Catholics, North and South.

It is probable that such a hybrid institutional settlement would be accepted by a clear majority of the Northern Ireland electorate, as well as by the electorates of the South and Britain, provided it was properly explained and sold by both governments. But most public discourse, as if mesmerised by the nation-state ideal, poses just two options, an exclusively British Northern Ireland, and its mirror-image of exclusive Irish sovereignty. However, Democratic Dialogue has proposed a referendum on three options; and, as Wilson suggests, 'a shared, pluralist Northern Ireland, linked to both the UK and the Republic' would have a good chance of getting most support even in a first-choice referendum – a majority of Catholics plus a sizeable minority of Protestants would clearly defeat the other two main options of 'Irish unity' and 'fuller integration into the UK' (Wilson 1996: 63).

The latter integrationist option, vigorously canvassed by so-called 'new (or liberal) unionists' such as Robert McCartney and Conor Cruise O'Brien in the UK Unionist Party (which depends on the support of the decidedly 'old' unionism of the Revd Ian Paisley's Democratic Unionist Party), runs counter to Britain's consistent strategy since early this century of keeping the North and its problems at arm's length. Even more relevant in the present context, it would greatly exacerbate the winner-takes-all non-solution of Northern Ireland's exclusive Britishness. In contrast, popular legitimation of the North being linked both to Britain and the South would be tantamount to a settlement. The traditional nationalist goal of a united Ireland would not be satisfied, but then Sinn Féin strategists have clearly lowered their sights from a unitary Irish state. As Percival (1996: 59) argues, they 'now see national self-determination as a process rather than as a prescribed outcome or solution'.

Yet the leaders of mainstream unionism refuse to recognise this shift. They resist any institutional expression of Northern nationalist identity, and the separate economic imperative for North–South institutions. For Ulster Unionist Party leader David Trimble, who insists 'my nationality is a zero-sum issue', this would diminish his own Britishness and make Northern Ireland a condominium ruled over by London and Dublin. This, however, is a misrepresentation, whether intentional or otherwise. North–South institutions do not have to mean London–Dublin rule. On the contrary, they would centrally involve the representatives of the North itself (who currently under Direct Rule from London have no executive powers even within Northern Ireland). Representatives from both North and South would gain a role in the common concerns of people in both parts of Ireland. Rather than being a one-way street, North–South institutions would express reciprocal linkages.

What is proposed is not a London–Dublin condominium, but continuing unionist intransigence could perversely lead to just that (as overplaying a weak hand has in the past led to unionism damaging its own cause). If mainstream unionism now persists in refusing to compromise on exclusive sovereignty, the two governments will have no other viable option but to move towards joint authority over the North. It might be argued that in any hybrid North–South settlement it is really only unionism which has to compromise, losing the actuality of exclusive British sovereignty, whereas Irish nationalism only has to give up an aspiration to its ideal. But the actuality is that unionists have no effective power, and the Southern government at present has an unreciprocated role in Northern affairs. It is only the perversity of zero-sum thinking which can see a gain in power and reciprocity as a compromise or loss.

Third time lucky?

Contemporary circumstances are much more favourable for border-straddling institutions than when the two previous attempts were made to defuse the border issue – the Council of Ireland which was to be part of the partition plan in the early 1920s, and the similar Sunningdale scheme of 1973–4. In 1920, the wider context was not Europe but the British Empire, while the nationalist ideal was enjoying its finest hour as Woodrow Wilson presided over the creation of a battery of new nation-states from the wreckage of the central European empires defeated in World War I (though Ireland was excluded as the British Empire was a victor). In 1974 the two states had only just joined the European Community, and the SEM did not exist. Today, in contrast, the combination of intensified globalisation, European integration and the fear of perpetuating an unprecedented twenty-five years of deadlocked military conflict might perhaps make it third time lucky.

In the 1920s the Council of Ireland was widely seen as simply a transitional stage on the road to a reunited Ireland. Resisting that outcome, unionists damned the Council as an unstable half-way house. But that would not necessarily be the outcome today, contrary to the unionist recycling of 1920s arguments. The EU itself already comprises a variety of well-established hybrid political forms; and just as shared sovereignty in the EU should not be seen as transitional to a single Euro superstate, North–South institutions need be neither transitional nor unstable.

> At the time of Sunningdale, one SDLP figure unfortunately described the proposed Council of Ireland as 'the vehicle which will trundle us into an Irish republic'. Yet today the scenario opens up, in a way that was hardly conceivable in 1973, that, instead of an either/or choice, Northern Ireland could both remain linked to Britain and become equally linked to the Republic.
>
> (Wilson 1996: 57)

The counter-argument that such a constitutional arrangement is not found anywhere else in the developed world, and therefore would not work, is conservative and deterministic, suggesting that because something does not already have an empirical existence it cannot be created. Of course, we should not underestimate the problems of North–South institutions, or their radical implications for two jurisdictions traditionally wedded to conventional sovereignty and the 'inside/outside' dichotomy between internal and external affairs (Walker 1993). A substantial amount of creativity and novelty is required if Ireland is to escape its national conflict and, in the process, pioneer a type of settlement of use elsewhere.

However, a major reason why North–South institutions have not yet begun to be initiated is government mismanagement, and unless policies are changed, the already fragile chances of a settlement could unravel. First, the Framework Document, far from being properly sold, was played down by the British government as unionists objected to it. Whether intentionally or through ineptitude, both governments allowed it to be virtually forgotten as soon as it was published in February 1995, their agenda dominated by fruitless posturing about the unattainable objective of prior decommissioning by the paramilitaries. The IRA cease-fire ended a year later. Second, the tactics adopted by both governments of giving reassurances that North–South institutions would not affect sovereignty is implausible and counter-productive. The whole point of such institutions is precisely to overcome some of the problems of sovereignty as currently understood, so either they would affect sovereignty, or if they did not they would be completely ineffective. As we have already seen, the absolutist conception of British sovereignty has already been breached in several ways by much less substantial developments. Third, both governments' strategies are overly reliant on the political parties (with their collusion, not surprisingly). They proceed as if Northern Ireland was a normally functioning representative democracy whereas the reality is Direct Rule, a party system effectively excluded from power and responsibility, unelected quangoes in unusually high numbers – even by British standards – and continuing paramilitary activity. Political parties are not the only, or in these circumstances necessarily the best, conduits of popular aspirations or willingness to agree. On the contrary, relying on the parties in the first instance, and involving the electorate only to endorse (or reject) what they eventually agree, is pathetically flawed. All the precedents of talks about talks suggest that if the process involves only the parties, it will not get beyond 'the first instance'.

Civil society, class and gender

Although the time may be ripe for North–South institutions, making them happen will require much more proactive, open and imaginative government policies, and participatory as well as representative democracy. Mainstream unionist appeals to democracy and majority wishes in Northern Ireland are disingenuous when the core problem is disagreement over Northern Ireland with its built-in unionist majority as the only framework for democracy. In contradistinction to Northern majoritarianism, democracy can only be enhanced by publicly accountable North–South bodies answerable both to the Dáil and to a Belfast assembly. And democracy is much more likely to be secured if participation in North–South institutions directly involves civil society.

Such a strategy would build on the fact that civil society in Ireland, North and South, is relatively transnationalised already, with both a heavy reliance on multinational capital and the extensive diasporas in Britain, North America and elsewhere. Furthermore, the North–South border is one of the most porous state frontiers in the EU. Partition was always partial, with many cultural and economic linkages remaining unbroken (in territoriality terms, many activities were not rebundled), and there is extensive – though uneven – social interpenetration across the border, despite the paucity of links at state level. There is indeed a glaring disjuncture between the flexibility of North–South links in civil society and the rigid views of territorial sovereignty, which have historically divided the state jurisdictions and driven a wedge of non-communication between the neighbouring administrations. Although far from being an absolute barrier, the border continues to be an obstacle to information flows and activity patterns in many fields.

The ensuing problems would be ameliorated by a North–South political framework, including more localised cross-border bodies within the border region. There are now a growing number of North–South community, voluntary and campaigning networks, business groups and trade unions, some of which are calling for more popular control over emerging North–South policies. To some limited extent, an informal North–South framework is already being constructed from the bottom up, and does not necessarily have to wait for official state sanction. However, state help and co-operation are essential, and the various networks and interests in civil society could be given their own North–South institution with a direct input into shaping North–South policy. An 'Island Social Forum', perhaps modelled on the South's National Economic and Social Forum, has been floated (Anderson and Goodman 1996), followed by a similar suggestion for an 'Irish citizens assembly' modelled on the assembly campaigning for home rule in Scotland (Percival 1996).

The South's National Economic and Social Forum draws up proposals on a wide range of social and economic issues, and includes representatives of women's organisations, environmental groups, the unemployed, young people, the elderly, the disabled, and other minority groups, as well as politicians, business people and trade unionists. An 'Island Social Forum', constructed on roughly similar lines, could help in democratising North–South relations and reconciling unionists and nationalists. It could make submissions to the executive North–South institutions, composed of politicians, supplementing their formal accountability with participatory democracy.

Although confined to the North, the Women's Coalition, constructed to represent women's groups, both unionist and nationalist, in Northern Ireland's 1996 election and peace talks, is indicative of the emerging possibilities. Rather than relying entirely on party politicians elected every four or five years, participatory democracy could involve a wide variety of organisations, some of which are themselves open to continuous democratic pressures. Many people, perhaps younger people especially, are alienated from conventional party politics but are nevertheless politically active in civil society. Furthermore, a reorientation towards civil society would fit in with the growing importance of the EU. Whereas lobbying Washington to get American investment or influence British policy both appeals to a largely external power and confirms the clientelism of party politicians, the EU and SEM are, as we have seen, substantially internalised. They offer more possibilities for popular, civic society involvement, for instance in the managing of some cross-border funds. EU officials claim that their funding mechanisms contribute to political development and reconciliation by promoting local dialogue (Wilson 1996: 71).

North–South participatory democracy, if instituted as part of the search for a settlement rather than awaiting the eventual outcome of party negotiations, would help to prevent party leaders or governments stalling or wrecking the construction of a settlement. It would enable the institutions of civil society to give the politicians some much-needed guidance. Part of the impact on the national conflict would, however, be more indirect, though that may be just what is needed. Political space would be opened up for mobilising around *non*-national identities, interests and practices, of the sort discussed elsewhere in this book, which span the North's sectarian divide and the border. Examples include social class, feminism, environmentalism, sexuality and other issues which have to varying degrees been crowded out by the national conflict – for nationalisms are often at the very least half-hearted about other movements or issues which might divide the national community or lead to some of its members fraternising with members of the other community. A North–South institutional framework could help a wide variety of groups, including ones whose interests and motives are quite separate from the twin dynamics of the SEM and parity of esteem. Conversely, this would help to move the political focus away from the fixation with national territoriality, thereby making a serious contribution to defusing the national conflict.

However, the notion that class, gender or other concerns could displace or replace the national issue should be rejected. This mistaken strategy was codified in Stalinism as the 'stages theory' and adopted by the Official IRA in the 1960s and 1970s: the first stage is to unite nationalist and unionist workers on a purely class basis while the question of nation is dealt with at a later stage. The problem is that – as Chapter 5 has shown – there is no pure class basis, and the national question cannot be side-stepped or put off to a later date. Parties in Northern Ireland which attempted to do that – explicitly Official Sinn Féin, implicitly the Northern Ireland Labour Party – either failed to get support or, in the case of the latter, quickly lost it when the national issue came to dominate the political agenda in the late 1960s. But we should also reject the mirror-image 'stages theory' of 'nation first, class later', which has epitomised some of the practice if not the rhetoric of Provisional Sinn Féin. Issues of class or gender cannot be put off to the promised land of a united Ireland, and attempts to do so can be guaranteed to leave many workers and feminists unimpressed. Conversely, the national issue cannot be solved by simply concentrating on it to the exclusion of other major sources of identity and material interest. There will always be tensions between class and nation, and tactical questions and disagreements about the relative weighting each should be given in particular circumstances. But both have to be fully taken into account. Whatever the leadership, it is mainly working-class people who do the fighting and suffering in national conflicts; as the United Irish leader, Henry Joy McCracken, put it, 'the rich always betray the poor', although he himself was one of the exceptions which prove the rule.

CONCLUSION

North–South institutions offer a route out of the dead-end conflict over territorial sovereignty, an escape from a parity of poverty and dis-esteem which is the reality for many Catholics and Protestants. The fixation with exclusive territoriality feeds zero-sum thinking and negative-sum practice. It misleads people into believing that there are just two options, an independent all-Ireland Republic or a purely British Northern Ireland, when in reality these are no more than mutually unattainable bargaining positions. In contrast, North–South institutions would provide a practical alternative to these traditional goals of unionism and nationalism and their opposing versions of the nation-state ideal. By bridging the border, they would puncture the pretensions of exclusive territoriality in line with the realities of the EU and the SEM. Enhanced by something like an 'Island Social Forum', they would help to meet the threat of economic peripheralisation in the SEM, facilitate the cross-border links of social, community and campaigning groups, and democratise the growing connections between the two parts of the island. Participatory democracy and a North–South framework would create much more scope for non- and anti-sectarian modes of political mobilisation based primarily on class, gender and other concerns which straddle both the border and the sectarian divide. And this, in turn, would help to diffuse the deadlocked national conflict, transcending its zero- or negative-sum terms. In a cumulative, mutually reinforcing process, it would further boost the so-called normal politics of other identities and interests, though these will continue to be subverted or downgraded until the national conflict is solved. The vicious circle of sectarianism could be replaced by a virtuous spiral.

The SEM implies a single island economy which in turn requires some island-wide institutions of governance. Together with parity of esteem in the North, this implies moving towards a more unified political culture and an all-Ireland society. In time, a North–South framework in the context of the EU would help foster genuinely all-Ireland identities and a stronger European identity as well, for identities are moulded in part by the institutional setting. Paradoxically, this could also result in a more unified Northern Irish identity, instead of the competing and largely sectarianised identities separated by the sovereignty issue. European integration is already bringing about some real changes in sovereignty and in attitudes to it – and even relatively modest changes could prove crucial in this situation of long-standing impasse. It may seem that the Irish conflict is immune to outside developments, as in Churchill's oft-repeated quote about 'the dreary steeples of Fermanagh and Tyrone' emerging unaffected from the mist after the cataclysm of World War I. But the EU and the SEM are no longer external developments – they are already internalised in the North, the South and the interrelationships between them.

Hence the inside/outside dichotomy simply reflects the traditional view of sovereignty and the outmoded debate about internal versus external solutions, as if it is possible to have one without the other. A so-called internal settlement within Northern Ireland would simply preserve intact the framework that is at issue, bottling up the sectarian conflict as if in a pressure-cooker. Confining the decisions on a settlement simply to the contested unit with its built-in unionist majority (or, even worse, to its party system where unionists dominate) is virtually guaranteed to produce stalemate, for the negotiating strategy of mainstream unionism seems to be to put off negotiating. A disinterested observer might conclude that both parts of Ireland, and indeed all three electorates of Ireland and Britain, have a democratic right to be involved in deciding any settlement as all are affected in various ways by the conflict. In turn, such a settlement would be more likely if the decision-making framework included all three, together with vigorous and imaginative joint strategy by the two governments.

As we have seen, the slow progress on a settlement reflects government mismanagement as much as the structural constraints. If that progress is reversed, which is always a danger, the main responsibility will lie with the two state administrations. However, provided there is better political management and more democratic involvement – a big proviso – Ireland could be a harbinger of transnational solutions to national conflicts. In a paradox of uneven development, where last can become first and there is such a thing as the privilege of backwardness, Ireland just might pioneer new hybrid types of institutions which could cope with the contradictions between national territoriality and transnationalisation. What is widely seen as a very traditional, old-fashioned, backward-looking conflict could conceivably give rise to political forms which would be very advanced, even in EU terms. It just might produce a viable alternative to chasing the elusive and tragic nation-state ideal. Let us hope that this is so, if only to spare us from yet more dreary repetition about Irish steeples continuing to poke from the mist, no matter what happens elsewhere in the world.

ACKNOWLEDGEMENTS

The chapter has benefited from joint work with James Goodman, and also with Ian Shuttleworth. Both commented on an earlier draft, as did Liam O'Dowd and Douglas Hamilton. My thanks to all of them, but the outcome is my responsibility.

REFERENCES

Anderson, J. (1986) 'Nationalism and geography', in J. Anderson (ed.) *The Rise of the Modern State*, Brighton: Harvester Press.

—— (1994) 'Problems of inter-state economic integration: Northern Ireland and the Irish Republic in the European Community', *Political Geography* 13, 1: 53–72.

—— (1995) 'The exaggerated death of the nation-state', in J. Anderson, C. Brook and A. Cochrane (eds) *A Global World? Re-ordering Political Space*, Oxford: Oxford University Press.

—— (1996) 'The shifting stage of politics: new medieval and postmodern territorialities?', *Environment and Planning A: Society and Space* 14, 2: 133–53.

Anderson, J. and Goodman, J. (1994) 'European and Irish integration: contradictions of regionalism and nationalism', *European Urban and Regional Studies* 1, 1: 49–62.

—— (1995) 'Regions, states and the European Union: modernist reaction or postmodern adaptation?', *Review of International Political Economy* 2, 4: 600–31.

—— (1996) 'Border crossings', *Fortnight*, 350, May: 16–17.

Anderson, J. and Shuttleworth, I. (1994) 'Sectarian readings of sectarianism: interpreting the Northern Ireland Census', *The Irish Review* 16: 74–93.

Breen, R. (1996) 'Who wants a United Ireland? Constitutional preferences among Catholics and Protestants', in R. Breen, P. Devine and L. Dowds (eds) *Social Attitudes in Northern Ireland*, VI, Belfast: Appletree Press.

Bull, H. (1977) *The Anarchical Society*, London: Macmillan.

Cadogan Group (1992) *Northern Limits: Boundaries of the Attainable in Northern Ireland Politics*, Belfast: Cadogan Group.

Colley, L. (1992) *Britons: Forging the Nation 1707–1837*, London: Yale University Press.

Harvey, D. (1989) *The Condition of Postmodernity*, Oxford: Blackwell.

HMSO (1995) *Frameworks for the Future*, Belfast: HMSO.

Kearney, R. (1988) 'Introduction: thinking otherwise', in R. Kearney (ed.) *Across the Frontiers: Ireland in the 1990s*, Dublin: Wolfhound Press.

McGrew, A. (1995) 'World order and political space', in J. Anderson, C. Brook and A. Cochrane (eds) *A Global World? Re-ordering Political Space*, Oxford: Oxford University Press.

O'Connor, F. (1993) *In Search of a State: Catholics in Northern Ireland*, Belfast: Blackstaff.

Percival, R. (1996) 'Towards a grassroots peace process', *The Irish Reporter* 22 June: 59–62.

Ruggie, J. (1993) 'Territoriality and beyond: problematizing modernity in international relations', *International Organisation* 47, 1: 139–74.

Sack, R. D. (1986) *Human Territoriality: Its Theory and History*, Cambridge: Cambridge University Press.

Trew, K. (1996) 'National identity', in R. Breen, P. Devine and L. Dowds (eds) *Social Attitudes in Northern Ireland*, VI, Belfast: Appletree Press.

Walker, R. B. J. (1993) *Inside/Outside: International Relations as Political Theory*, Cambridge: Cambridge University Press.

Wilson, R. (1996) *Reconstituting Politics*, Report 3, March 1996, Belfast: Democratic Dialogue.

INDEX